Be Your Own House Contractor

5th Edition

Be Your Own HOUSE CONTRACTOR

Save 25% without Lifting a Hammer

Carl Heldmann

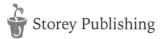

Storey Publishing

The mission of Storey Publishing is to serve our customers by publishing practical information that encourages personal independence in harmony with the environment.

Edited by Lisa H. Hiley
Art direction and text design by Cynthia McFarland
Cover design by Kent Lew
Text production by Kristy MacWilliams
Indexed by Chris Lindemer
Photography by Carl Heldmann

Printed in the United States by Versa Press
20 19 18 17 16 15 14 13 12 11

Library of Congress Cataloging-in-Publication Data

Heldmann, Carl.
    Be your own house contractor : save 25% without lifting a hammer / by Carl
    Heldmann. — 5th ed.
        p. cm.
    Includes index.
    ISBN 978-1-58017-840-2 (alk. paper)
    1. House construction — Amateurs' manuals. 2. Building — Superintendence — Amateurs'
    manuals. 3. Contractors — Selection and appointment — Amateurs' manuals. I. Title.
TH4812.H45 2006
690'.837 — dc22   2005036253

Contents

Preface to the 5th Edition

After 25 years in print helping many thousands of happy readers successfully build their own homes, what could be exciting enough for a new edition of this book? One significant addition is a Web site extension, www.byoh.com, where you'll find:

◇ Downloadable software to control building costs

◇ Up-to-date, average "cost to build" figures for the U.S.

◇ Current mortgage rates, updated daily

◇ Newsletter with services, product reviews, new technology, and letters from other homebuilders

◇ Links to many useful sites with owner/builder construction lenders, house plans, site supervisors, and other valuable information

In addition, the chapter on cost estimating, the most important part of building, has been greatly expanded and the chapter on construction financing has been extensively rewritten to reflect major changes in that industry. We have also added a practical and useful English/Spanish glossary.

I hope you enjoy this 5th edition of my book, my new Web site, and most of all, your own building experience. I hope you save more than 25 percent and have fun! Let me and your fellow owner/builders know how things go, via the www.byoh.com newsletter. We care.

Acknowledgments

Without the help of my wife, Jane Prante Heldmann, and her mother, Clella Hunt Prante, the original book would never have been written. I would like to thank my editor, Lisa Hiley, for her help with this edition.

Introduction

THE FOUR QUESTIONS I HEAR MOST FREQUENTLY in my business are:

1. What is the actual cost to build a house?
   Answer: About 75 percent of what new homes sell for in your area.
2. What does "contract my own house" mean?
   Answer: That you are in charge! You're the CEO.
3. Do I have to be licensed?
   Answer: Only if you want to become a professional builder.
4. What is the number one problem that builders run into?
   Answer: Cost overruns.

And the one other question that I hear all the time is, "Can I really save 25 percent without lifting a hammer?" I know from my own experience that you can. Here's how it all began.

Back in the 70's, a friend who was a home designer suggested that I build my own home, as we couldn't afford the dream house we wanted. I told him he was crazy, that I didn't know anything about building. Besides, I had heard that builders make less than 10 percent profit.

"Relax," he said, and proceeded to explain to me that merely by being the general contractor (GC) and hiring expert subcontractors to do all the work, I would save 25 percent or more on the cost of my dream home. I was skeptical, but I tried it. I saved more than 25 percent, and it was so easy that I changed careers. I obtained my builders license and became a professional builder. The rest, as they say, is history.

I wrote this book in 1982 and recently built the accompanying Web site (www.byoh.com) so that you'll have an easier job building your first house. You won't have to make the same mistakes other builders and I have made. With this book and Web site as guides, not only will you save money and avoid mistakes, but by being your own "boss" of the building job, you'll get exactly the house you want (based on your budget, of course), done the way you want it done.

Over the past 30 years, I've been involved with thousands (yes, 1,000s) of owner/builder projects and have seen some people save more than 25 percent. It all depended on how well they shopped and how well they controlled the costs. Find out how in chapter 3. (By the way, if you want to save a little more money, you can even do some of the labor yourself, if you know what you're doing. If you don't, just settle for a 25 percent savings as the GC.

Just remember, every builder in the world started with their first house. Most of us didn't have a book or a Web site to guide us. You do. However, this is not a technical book on how to build a house. It won't teach you about wiring or plumbing, because you truly don't need to know. It will teach you how to find, hire, supervise, and pay the professional electricians or plumbers who do know how, and at builder's cost. That's the role of the GC.

Here's a true story. One morning a number of years ago, I received a phone call that went as follows:

"Is this Carl Heldmann?"

"Yes, it is."

"Are you the one who says it's easy to build your own home?"

By now I was a little nervous and expected to hear the worst. But I said, "I sure am!"

"Well," she replied, "I just wanted to tell you that it's even easier than you said it was!"

I breathed an audible sigh of relief.

The caller went on to say, "My husband and I just finished building our own home in Windy Hill, and we saved so much money and had so much fun that we bought the lot next door and we're going to do it again!"

Imagine, they had so much fun they wanted to do it again! I hope your experience is just as enjoyable. It might even lead to a new career.

Chapter 1

Be Your Own General Contractor and Save

THIS BOOK WILL TEACH YOU how to be a general contractor for building your own house. There's one big reason for doing this — to save money. How much you save will vary considerably depending on local prices for labor and materials, land costs, building permits, the size of the house, and your ability to follow the steps outlined in this book.

If you do all that I suggest, you can save as much as 25 percent of the total market value of a house. If you think of how much you will have to earn — and pay taxes on — to accumulate the money to build, the savings are even greater. And by reducing the amount of money you will need to borrow and then pay back with interest, you will save even more money for years to come.

Let's look at what you can expect to save. Your savings are based on the difference between what the house would sell for when finished (the market value) and what it cost you to build.

The size of the house will be the largest determining factor, as most general contractors base their profit and overhead on a percentage of the total cost of the project. A larger house costs more and therefore will include a larger profit and greater overhead for the builder. You will also save more in real estate commissions as the size and value of the house increases.

You can determine the market value of your new home prior to making a final decision about building. Here's how. Once you select land, obtain your house plans, and complete a list of all the items you plan to put into your house, such as flooring and cabinets, a licensed fee appraiser can determine the fair market value of your future house. Your lender will order an appraisal as part of the loan process, so you could wait until then to make a final decision. If the appraisal shows that you aren't saving enough money based on your cost estimate, there's probably an error in your cost estimating efforts.

Say that as a contractor, I build a house that I am offering for sale to the general public. My costs would typically break down as follows: the land usually costs 25 percent of the selling price, with labor and materials taking another 50 percent. That gives me a gross profit of 25 percent.

Wow, you say. That's a lot of money to make off of one product. Well, if it were that simple, you would be right. But before you get outraged with the building industry in general, let me show you where that gross profit goes when you build professionally.

First, I have to pay related sales expense out of my gross profit. If that involves a commission to a real estate broker, it may cost as much as 6 or 7 percent of the selling price. Next, like any business owner, I have overhead expenses. This varies with each builder, but the National Association of Home Builders suggests that homebuilders allocate 50 percent of their adjusted gross profit (after sales expense) for overhead expenses. These include, but are not limited to, phone, insurance, secretarial, transportation, rent, and office equipment. That leaves me with just 9 percent, which is half of the *adjusted* gross of 18 percent (25 percent minus 7 percent real estate commission). That is my real expected net (before taxes) profit. You can see that when builders say they make less than 10 percent, they're not wrong.

But you aren't building professionally. You don't have to pay sales expense. You don't have to worry about business-related overhead. You can take the entire gross profit of 25 percent and consider it yours. You may never even have to pay taxes on that gross profit if you follow the Internal Revenue Service guidelines for reinvesting your primary residence capital gains (check with your tax advisor). Imagine! You could actually "make" this kind of money while you go about your regular daily business.

What You Need to Know

You need to know very little about the actual building process to be a general contractor. You don't need to have technical knowledge about framing or bricklaying or wiring. Your subcontractors will know their business just as mine do. I'll help you make sure of that.

You may wish to pick up some information on various aspects of building, and that's fine. There are many excellent how-to books available for the do-it-yourselfer on almost all phases of construction. You may want to read some of them to better understand the process of building a home. But there's no way you can become a master of all trades. Your role as the general contractor is to be an organizer and a manager, not a tradesman. Your responsibility is to get the job done by other people.

If you can estimate your costs, control those costs, and deal with people in a fair manner, you can build your own house! The most difficult parts of the process will be behind you when you actually start construction. Sounds unbelievable, doesn't it? But it's true. Your job of planning, estimating costs, and organizing will be 90 percent complete when you break ground . . . or it should be. At that point it is up to your team of experts — your subcontractors (subs) and suppliers — to do their job. If you choose them carefully, they will do their jobs correctly, even without you being physically at the building site.

Even though you don't need to be an expert, remember that there are no silly questions. Never be too proud to ask questions at any point in the project from either a supplier or subcontractor. Most of them are very willing to help. They don't make any money until they sell you something or perform a service.

Don't look at building a house as one huge job. Viewing each phase or step as a separate job that you can easily accomplish and cross off your list makes the overall task seem less monumental.

Find a Good Carpentry Crew

Perhaps the most important step in building your own home, and maybe even in helping you make the decision to do so, is finding a good carpentry crew. Also known as a framing contractor or framer, the carpenter, along with his crew, is your key subcontractor. An experienced, reliable carpenter, hired at the beginning of the process, will help you make many intelligent decisions and will help your local lender make a decision to lend you money.

A personal visit to a few lumberyards or building supply houses (especially smaller, locally based ones) to ask for recommendations will give you more than enough leads to find a good carpentry crew. Asking friends and acquaintances who have been through the building process or a major renovation is another way to gather names of good carpenters (and to avoid bad ones). If the first carpenter you find is too busy, ask him to recommend another. But usually another house can be worked into the first carpenter's schedule, and it is worth waiting for a good crew. Your carpenter will be one of your best sources for finding many of your other subcontractors.

Keeping the Records

Unless you're an exceptionally orderly person, you'll find yourself with some construction papers at your desk at home, more in your car glove compartment, and others scattered in pockets or around your place of business.

Avoid this by getting organized early. Use a briefcase as your traveling office file. Buy a bunch of manila folders and label them as you need them. You'll need one for finances, another for inspections, one for each of the subs, and so on.

There are two reasons for doing this. First, it will keep your work flowing more smoothly if you don't have to paw through a jumble of papers to find what you want. Second, you'll develop a feeling of confidence just from seeing your part of the job arranged in such a businesslike way.

After selecting a good candidate for this job, but before hiring him, check him out. He should give you a list of four or five jobs he's handled. Talk to the owners. Look at the work he did for them. Other sources to check for references with are building suppliers that have dealt with him.

Most experienced carpenters have a thorough knowledge of most phases of construction. These job-site veterans are experts on getting a house framed up and closed in quickly, with all doorways, stairways, and window openings sized and located correctly. A good carpenter may be hard to find and hard to get, but is always worth waiting for. He may cost you a bit more up front, but in the long run will save you money and time.

Time Involved

How much time is involved in being your own contractor? This will differ for each individual, but you'll spend more hours planning and preparing than you will on the construction site — that is, if you let the subs do their jobs without feeling that you must be there every minute. (After all, does your boss stand and watch over you?)

You may spend a month or several months of your spare time in the planning stages, depending on how quickly you are able to make decisions on all the various aspects of the project. For example, settling on house plans can take an hour, a month, or even longer, and the same thing is true when you are choosing land, specifications, and subcontractors.

You won't need to spend a great deal of time on the project after construction starts. Your maximum time commitment ought not to exceed two hours a day, which need not interfere with your job or normal activities. You will have many helpers and will be able to take advantage of some free management by other people. For example, you can have the real estate agents (the ones you are dealing with on your lot) do some legwork by checking on restrictions and getting information on septic systems and wells. Your suppliers can also help you save time by finding some of your subs for you, preparing a list of the materials you need that they hope to sell you (this is called a takeoff), and giving technical advice.

Work by Phone

As a contractor, the success of your house-building venture depends on good communication. Subcontractors have cell phones and/or pagers, making communication a lot easier than it used to be. Frequent phone calls to subs are the key to this business. Make those calls before you leave for work or in the evening when you are not working. You can make on-site inspections before or after work or on your lunch hour.

Daily on-site inspections usually aren't necessary. If you believe they are, and you can't do it, ask your spouse or a friend to check in, but you don't have to be there every minute. There isn't a general contractor alive who is. That would mean that he could build only one house at a time, but most contractors have five or six homes under construction at once.

You don't have to watch masons lay every brick or your carpenters pound every nail. Allow them to do their jobs. Mistakes may be made from time to time. Chances are they would have been made if you were there. There isn't a mistake in the whole process that can't be rectified. It should be your individual decision on how much time you want to spend at the site.

Years ago I built a vacation home in the mountains about 140 miles from our permanent residence. I made only three trips to supervise and check on the progress the entire time the house was being built, from the staking of the lot to the final inside trim. While I don't recommend this, I mention it to show that you don't have to be there every minute or every day. My mountain house was built by phone. This was far cheaper and much less tiring than driving 140 miles each way. The subs did a beautiful job. They wouldn't have done any better if I had been there watching them.

You Can Get a Loan

Unless you are paying cash for your house, you will need a loan. Almost all loans for houses are made by banks, mortgage companies, and credit unions. Shop the Internet or use a mortgage broker. A good broker will knowledgeably research all sources for your specific loan needs, saving you time, money, and a lot of headaches.

Obtaining a loan is crucial. No money, no house. You must determine very early in your decision-making process whether you qualify for a loan.

The intricacies of financing are explained in chapter 4, where I discuss how a loan works and how much you can borrow. Lenders used to be somewhat reluctant to lend to an individual who plans to be his or her own general contractor. Not any more! But lenders do want to be sure the house will be built properly and, most important, finished. They don't want to step in and finish a house. They are lenders, not builders, and that's the way they would like to keep it. It will still be a challenging sales job to convince them you can be your own general contractor, but with the knowledge from this book, you can do it.

Plan your loan application process carefully. Before meeting with a loan officer, you need to have thought through all the steps in the construction of the house, to understand the problems, and have solutions to them. If the lender isn't enthusiastic about your decision to be your own contractor, you must do all in your power to present this approach as an asset, not a liability. Your projected savings can make it an asset.

The lender will want to know that you are an organized person who can handle the details of home construction. You will be convincing if you have all of your homework completed when you reach the bank.

Have It Your Way

A major benefit of acting as your own general contractor is that you will get more of what you want in your house, with fewer hassles. Most professional contractors (builders) are stubborn. They tend to like to do things the same old way. This often causes problems for the buyer who wants something done a little differently. With you as the general contractor, you can have things done as differently as you like.

Be Persistent!

After you become familiar with the material in this book, there should be no reason for you not to get a loan if you have good credit. It costs very little to go through the initial stages of discussion with a lender and get a commitment for a loan. Keep in mind, however, that one or two lenders may turn you down before you get an approval. Don't be discouraged. Sometimes it has nothing to do with your ability. Sometimes you pick a lender who is not interested in your situation. That's just the way the lending industry is.

In building my first house, I was turned down by four lending institutions before I found two who would permit me to be my own general contractor. If I hadn't been persistent, who knows where I would be right now? I was turned down because I wasn't thoroughly prepared for the loan application. Each time I was rejected, I would do a little more homework and be better prepared for the next appointment. I didn't have a book to guide me — you do. You can do it.

Three Alternatives to Being Your Own House Contractor

Let's suppose that for one reason or another you (or a lender) decide you can't go forward with your plan to act as your own contractor. Here are three other ways you can build, each using a professional general contractor (builder) in a position of increasing responsibility and at correspondingly greater cost to you. (The first two options should still save you money overall.)

1. A manager's or site supervisor's contract
2. Cost plus a percentage, or a fixed fee contract
3. Contract bid

Because each option increases the contractor's responsibility, the cost of using that contractor also increases. The cost of the land, the materials to build the house, and other fixed expenses should remain the same.

For all practical purposes, options 2 and 3 are readily available contracts that are used by most builders with varying degrees of legalese and varying degrees of slant favoring one or the other party. (Samples of contracts for all three options are shown in the appendix on pages 116–125.)

The least expensive way to go — and the one by which you can still be considered the general contractor, be accepted by your lender, and possibly feel more comfortable dealing with subs — is the first option, the manager's or site supervisor's contract.

Under this arrangement, you hire a licensed general contractor whose one responsibility is to act as your manager with the subs. This contractor, in return for about one-third of the normal fee or profit and overhead, will assist in finding the subs (although you can still find your own), schedule the subs, check the quality of their work, approve the quality of materials, and order materials, when needed, in your name. You will still be responsible for selecting and buying the land; figuring out all cost estimates; securing suppliers, permits, and loans; paying all bills, including those from subs; and ensuring inspections for quality and approval. You will be responsible for the final job and its overall acceptance.

One advantage of this contract is that the general contractor will work with you in the following ways:

◇ Assist in cost estimating

◇ Help find subs

◇ Schedule your time better

◇ Help cut red tape and get the necessary permits

◇ Schedule subs, from survey to landscaping

◇ Assist in reviewing bills when requested

The manager's or site supervisor's contract shown in the appendix can be modified to include more or fewer of the responsibilities of the general contractor. The cost, of course, will vary with the number of responsibilities listed.

If you decide to use any of the sample contracts provided, you or an attorney can modify it for applicability and to make certain it conforms to local laws.

Chapter 2

Where to Start

THESE ARE THE BASIC STEPS in deciding whether to proceed with your building project.

1. Make a budget.
2. Deduct the cost of the land.
3. Determine what size and style of house you can afford.
4. Find house plans that meet your criteria.
5. Determine the cost to build this particular house.
6. Get an appraisal of the finished house and land together.
7. Now you can make your decision.

The Budget Comes First

Making a budget *has* to be your first step, even if you are a billionaire! Making a budget for housing is simple. Don't complicate it. The formula is: Cash + Loan = Budget!

First, add up all your available cash and any assets you want to convert to cash for your dream home. Second, determine your maximum "borrowing power." These two steps determine how much you can spend. It's that simple.

The first step needs no explanation. The second is easy, too. Just contact any mortgage lender, and a loan officer will look at your credit history, cash available, your income(s) from all sources, and your total monthly obligations (debt) to determine the maximum loan amount you will qualify for based on current interest rates and loan terms (length of loan) available. It makes sense to contact several different lenders to compare rates and terms.

Land Is the Second Step

You only need to select the land at this point. If you find a particular lot and don't want to risk losing it, you can make an offer with the understanding that your binder or deposit will be refunded if certain conditions or criteria are not in your favor or if you can't obtain construction financing. This is called a contingency offer. Don't put more than 10 percent down as a binder or deposit, and if you are using a broker, be sure that the money is held in escrow.

If you already own land, jump ahead to the next topic. If not, read on. If you don't already own land, you must decide where and what to buy. The cost of the land should be your guide, for it will be the first item subtracted from your budget. Whatever is left in your budget after subtracting the cost of land is *all* you have left to build the house. Relax! The cost of the land (or at least most of it) will be included in your construction loan, as you will see.

Choosing the Right Land

First choose the *area* where you want to live, for location is the main determining factor in land cost. Next, look for a lot or a site with the acreage you want to accommodate your personal tastes.

If you have lived in a city or town for a while, you probably know where you want to build. If you haven't been looking around, or if you are new to an area, I suggest working with a local real estate broker. These brokers know what each neighborhood offers and what lots cost in different locations. If acreage is what you are after, here, too, a real estate broker is most helpful. A broker can help you locate the property you want and can help with all the details necessary to assure you that it is a suitable building site. The broker should be able to show you a survey (map) of the lot and point out the boundaries to you during a walk around the lot.

I can't say this strongly enough: Find the land first and then make the plans to fit it.

Do You Need a Basement?

Do you want — or need — a basement? Here are the pros and cons of building a house with a basement.

Pros:

◇ It's relatively inexpensive, compared with aboveground space, particularly if you must dig down 6 or more feet anyway to place footings and foundation walls, as you must in some parts of the country.

◇ A basement locates your heating system, and sometimes the fuel supply, out of the way.

◇ It offers extra storage space, though it may be too damp for paper, metal, and clothing.

◇ On a sloped lot, a basement may provide you with a different garage option.

◇ It's an excellent location for an out-of-the-way family room or den.

Cons:

◇ If the water table is high, flooding may be an annual headache.

◇ A basement may be too damp for many uses, or require a dehumidifier.

◇ It's an ideal spot for storing things better taken to the dump.

◇ Building a basement can be expensive, particularly if the excavator encounters ledge (large rock mass) that requires blasting or a spring that keeps flooding the site.

Sloping Lots

If you want a basement and you live in an area of the country where the soil doesn't drain well, avoid flat lots. A sloping lot will provide drainage if you install footing drains. Another advantage of a sloping lot is that it allows you to design your house with a walkout basement with doorways or windows on one side to provide natural light and ventilation. Also, in some areas the open side can be frame construction, which is a little less expensive than poured concrete or concrete block.

If you have no alternative to building a basement on a flat lot with poor drainage, be sure to hire a professional waterproofing subcontractor. He will take extra care in waterproofing the exterior basement walls, providing footing drains that drain either to a sump pump or to a lower elevation on your property. Be sure to follow local codes regulating where the drainage can be discharged.

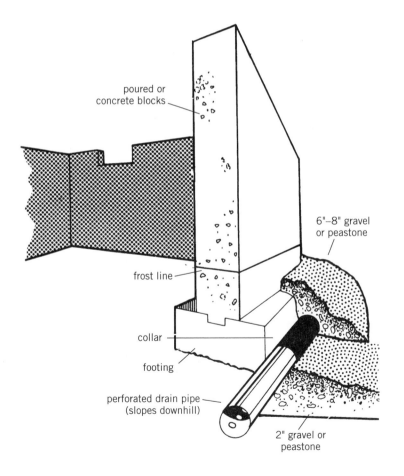

poured or concrete blocks

6"–8" gravel or peastone

frost line

collar

footing

perforated drain pipe (slopes downhill)

2" gravel or peastone

A footing drain will help to keep your cellar dry.

If you don't want a basement, try to find a relatively flat lot so that you won't have an excessive amount of crawl space (or fill, if you decide on a slab foundation). Building a crawl space is cheaper than digging a basement. If the lot slopes on only one end or corner, though, it will not cost too much more to accommodate the foundation to the lot. Notice that you are accommodating the foundation to the lot, not the other way around.

Other Considerations

Trees are valuable; if you like them, try to find a wooded lot within your budget. Barren lots are cheaper to build on but more costly to landscape. A wooded lot generally costs more to buy and more to build on, but less to landscape. It's almost an exact cost trade-off. It's your choice, but if you are concerned about top-dollar resale in the near future, it's worth putting a few thousand dollars extra into a wooded lot with mature trees.

If the area in which you are looking has no development activity near it, I strongly recommend taking test borings of the soil before you purchase the lot to determine its load-bearing capability. These tests will also show whether there is ledge on the site that might require blasting. The test is not expensive, and the seller should be willing to pay for it. Without a soil test, you can end up paying thousands of extra dollars for foundation and drainage work. Be sure that this test is included as a contingency in any contract to buy. It can also be included as a refund provision in the contract in the event that non-load-bearing soil is discovered after purchase.

Specialists in test boring for load bearing are listed in the Yellow Pages under "Engineers, Consulting" or "Engineers, Foundation." Most county health departments make soil tests for septic systems. This is done for free or at a very low cost. These departments also provide information about wells.

Let the Brokers Do the Work

Put the burden of arranging these tests on your real estate brokers. Have them do the legwork and scheduling. Just make sure those tasks are in the contract you sign with them. Let them earn their commissions by handing you a nice, clean, finished deal — a lot ready for you to build your dream house on.

Make sure you deal with a realtor who is a member of the National Association of Realtors so you have recourse if there is any problem. In almost all towns and cities, the local Board of Realtors is so image conscious and worried about a member ruining that image that it will protect you, the buyer. Simply call the board if you believe a realtor isn't doing a good job or isn't representing your best interests.

Building Permits, Zoning, and Title Insurance

Is it a buildable lot? This is the most important question to ask when deciding to buy a building site, and your local building inspection department is responsible for answering it in the form of a building permit. They will *only* issue a building permit if the property can be considered a buildable lot. Be sure to check with your local building inspection department before you purchase your site.

Consider zoning carefully when you are choosing a lot or acreage. Be aware of what could later be built in your future neighborhood — stores, offices, trailer parks, or industry. Look at what is already there. Such things as dumps, railroads, and industrial buildings can make a piece of land worth much less as a site for a home. One could write a book on zoning — and many have. A quick consultation with your real estate broker and/or the local zoning department should resolve any concerns you have.

If water and sewer are provided, be sure you check out all costs to get the water and sewer to your property line if it is not already there (it may be across the street). You also need to find out about any tap-in fees or privilege fees charged by the municipality or the association providing the services. Do the same for gas, electrical, and phone services. The process of thoroughly investigating wells, the septic system, load-bearing capabilities, water and sewer fees, and zoning may take a few weeks.

You also need to check out all restrictions on the size and type of dwelling you can build and what a neighbor may build. These restrictions may or may not be specifically covered in local zoning regulations. Different parts of the country use different means to protect an area from visual blight.

Finally, the seller of the land should always provide you with a title insurance policy, which insures that the property has a "clear title." The insurance company searches for previous transactions to guarantee the policy. As with any insurance policy, its purpose is to bring you peace of mind. If you want added protection in a land transaction, consult an attorney who specializes in real estate.

Checklist for Buying a Lot

Here are some of the many things you will want to check before buying a lot:

Building permit
Most important: Can you get one for the site?_____

Neighborhood
Will your house fit in with others on street?_____
Quality of schools? _____
Attractive street? _____

Transportation
Public transportation available? _____

Pollution (air, water, or noise, such as airport) _____

Zoning restrictions _____

Sewage, water available _____

Soil tests made _____

Direction of slope, if any _____

Drainage (Observe after heavy rain, if possible.) _____

Test borings _____

Size of lot _____

Trees (valuable, but sun needed for garden) _____

Title clear _____

Cost (A counteroffer of 10–20 percent below asking price
      is often accepted.) _____

How Much Should You Spend on Land?

Buying land is a very subjective process, and the cost of land varies greatly in different regions of the country. One tip, however, is to always keep resale value in mind. I know that may not seem important to you right now, when selling the property is the farthest thing from your mind, but some day it will be. Remember, resale is always on the mind of your mortgage lender.

I recommend spending no more than 25 percent of your budget on the land. While this is not always possible, and gets harder to do every year, it is an excellent guideline. If you have to spend more than that, something else has to give, and that will be the size and style of your dream house. Size and style determine building cost. (Size more than style.) You'll have to juggle hypothetical scenarios in the planning process until it all fits into your budget. In order to do this, you will have to know up front approximately how much it costs to build a house.

The actual "cost to build" anywhere in the United States is the most closely guarded secret in the industry, but with the Internet, it is easy to find out information on building costs in your area. You can also stay abreast of current mortgage rates online.

One method of "eyeballing" building costs is to find a new home being built by a professional builder that is for sale and that is similar in size and style (and quality) to your dream house and do the following: Take the sale price of that house and deduct the land cost, real estate commissions, and 25 percent builder profit and overhead and you'll have the *real* "cost to build." The land cost may be a bit tricky to determine, but any real estate agent can find out for you. You could even call the builder. All homes, even used homes, have the "site value" broken out separately. Now you simply divide the "cost to build" by the square footage of the heated area of the house and you'll have the cost per square foot.

If you are successful in keeping your land costs at 25 percent of budget, you will have 75 percent of your budget left for the cost of building the house. Take that number, divide by the "cost to build" and you will now know what size (in square footage) house you can afford. Now you can start looking at house plans by square footage, and, of course, style.

Water and Sewage

You will need to consider both water and sewage as you select your building lot.

If water and sewage are provided by your city or town, you need only find out how to tap into them, and what costs are involved.

If either one is provided (and quite often it is only water), you will not face much of a problem. You may need some type of a permit, plus a percolation (perc) test for a septic system, to determine whether the soil will absorb the discharge of the system.

If you must provide both water and sewage, some study of the situation is needed.

A septic system consists of a sewage pipe running from your home to an underground septic tank, usually made of concrete. Sewage enters the tank and is broken down by bacterial action. The resulting liquid then flows from the tank into a system of perforated pipes that distribute it into an area called a leach field. The size of the field required depends on the amount of

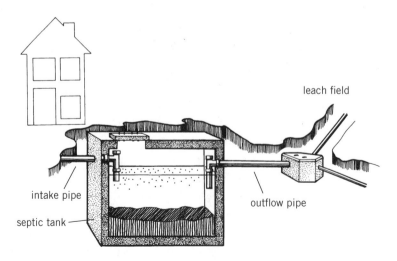

Typical layout for a septic system. The leach field must be located in an area where the soil can absorb moisture.

sewage going into the tank and the ability of the soil to absorb moisture. Sandy soil is best because of its permeability; denser, claylike soil can create problems.

Current regulations in many communities may require that new septic systems be designed by an engineer. In many areas, regulations require that a well be a certain distance, say a minimum of 150 feet, from your leach field or from the field of any neighbor. Thus it's best to have a large lot if you must provide both water and a sewage system.

Digging a well can be expensive. You can get a dowser, complete with forked branch, to "locate" water for you, or you can do what most well diggers do, which is to locate the well where it's most convenient and start digging.

Most well diggers charge by how far down they bore, and they usually can estimate about how far they will have to dig. The depth of existing wells in the area is sometimes (but not always) a good indication of how deep you'll have to drill to reach water, so try to use a well driller who is familiar with the area where your lot is located. Some diggers will offer you a contract price for the well rather than a per-foot rate.

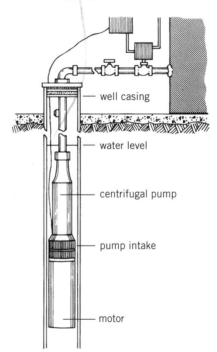

well casing

water level

centrifugal pump

pump intake

motor

A common type of installation in a home well. The pump and motor are positioned at the bottom of the well.

The House Plans

Rule of thumb: Size matters. The bigger the house, the more it costs. Here's a tip: Two-story construction is cheaper than one-story. Two of the most expensive parts of a house are the roof and the foundation. A two-story home with the same square footage as a one-story has half the roofing costs and half the foundation costs. There are also economies in plumbing and heating when building two stories.

Square footage refers to the heated (livable) area of a house, although it is determined by measuring from the outside surface of exterior walls. For example: a simple house that measures 40 feet x 30 feet = 1200 square feet (sq ft). No deduction is made for the thickness of walls.

There are many thousands of house plans available and several ways to find them. Some places to look are building or renovation magazines, books of plans, the Internet, or on CD-ROMs. There are even computer programs that allow you to design your own house plans, although these are not always as easy to use as the creators would lead you to believe. Another more expensive possibility is to have a local architect or draftsman/designer draw your plans.

The least expensive way to obtain plans is through the thousands of magazines, books, Web sites and CD-ROMs of plans that are available. Quite often you'll find a plan that requires only minor modifications to make it suitable for your needs and your building site.

You either should stick with the plans as drawn or ask advice from a draftsman/designer or an architect on any changes, no matter how minor. This will be much less expensive than having the entire plan drawn for you. An experienced draftsman can even make major changes in a set of plans and advise you on the practicality of changes you suggest and the additional cost, if any, of those changes.

If you do hire a draftsman/designer or an architect, obtain estimates from several before hiring one. Ask for examples of work done and call their references.

Make Changes Early

Try to make all changes in the plans before you start construction. You need not have the plans redrawn for minor changes such as moving, adding, or deleting a window or door, but you should if you move walls or change roof lines and roof pitches. On-the-job changes are expensive, so live with the decisions you make on paper or expect high cost overruns.

Pay strict attention to what is in the floor plans. Picture yourself walking through the house from room to room. If you are unsure as to whether a room is large enough, find a room of about the same size in another house, and compare.

This is a good time to mention the only two tools you'll need as a general contractor: a 25-foot tape measure and a cell phone.

Studying Plans

Figuring out the plans is easy, although it may look difficult at first. Look closely at room sizes, room placement, traffic flow, kitchen workflow, closet space, number of baths, and overall size. Those are the major functions of design. You can move, add, or delete doors and windows without too much effort — on paper.

Unless you have an unlimited budget, and very few of us do, a house is a series of compromises. We can't afford all of those beautiful things we see in home magazines. A house is going to cost so much per square foot using the average number of windows and doors and average appliances. If you want an expensive whirlpool

Sometimes plans are sold in sets; sometimes you can order them singly. You'll need about six sets of plans: one for yourself, one for your lender, and one for each of the major subs. It's good to have as many as possible. If you're going out for bids, for example, each of the bidders for the job should have a copy of the plans to study.

tub, you may have to compromise on something else, such as the size of the house, adding a garage, or installing a paved driveway.

Don't make any of your decisions a "matter of life and death." You may be surprised how unimportant that "big decision" becomes when you are finished. Agonizing over decisions only leads to friction — within yourself and with others. Make all of your decisions conscientiously but quickly, and then move on.

Be sure your plans include specifications (specs). Specifications are lists of the materials that are going into your house. Most plans come with a set of specs and include a "for input" section for you. Refer to the Description of Materials form on page 126 in the appendix, and you will see that there are places for indicating decorative items, such as wallpaper, carpet, paint, stain, and door hardware. Obviously, you will choose these. Since specs come early in the game, most people haven't selected these decorative items at that point. Therefore, monetary allowances are used with the term

Cutting Construction Costs

Your first chance to save a lot of money is now, when you are selecting the plans for your home. Think about the following:

◆ How many square feet do you really need? Every reduction you can make in size will cut the costs.

◆ One storey or two? It's cheaper to build the same number of square feet of housing in a two-storey house.

◆ What is its shape? A simple rectangle is the cheapest to build. Curves, extra corners, and extensions in the floor plan all increase costs.

◆ What style roof? The least expensive is the common gable roof. Dormers and multiple gables make the roof more difficult to build and therefore more costly.

◆ How's your timing? If you're building when there's a lull in construction (such as in winter), chances are you'll get your money's worth — and maybe a bit more.

Insulation

You're lucky. While those of us who own homes built before 1970 wish we had put more insulation in them at the time, you are starting fresh and can plan a home that is energy efficient.

Consider this as you select your plan and as you talk over your plans with your subs, particularly your carpenter and your insulation sub. Your choices here can result in thousands of dollars of savings as you live in your home over the years. Some decisions will require changes in your house plans and specifications.

If you have a full basement, should you insulate it? Do your windows provide two layers of glass? How will your outer walls be built? (Three possibilities are shown on page 26.) How much insulation will you have over your head? Will your heating ducts and hot-water pipes be insulated? Is your chimney inside the house, where any heat it collects will be returned to the house, or on an outer wall, where heat will be lost? Does your fireplace provide a source of outside air so that you're not sending heated air up the chimney, and does it have a tight damper so hot air isn't flowing up from it when it isn't in use?

A wood-burning stove, a pellet stove, or a gas fireplace can give you much of the ambience of a traditional fireplace while saving you hundreds of dollars in construction costs and hundreds more later by providing more efficient, less expensive heating.

Utility companies often discuss R factors. This refers to the ability of a material to resist the escape of heat. A high R factor means that you are not losing as much heat. Recommended minimum R factors for walls and roofs vary geographically, so check with your building inspector to see how much insulation is typically used in your area.

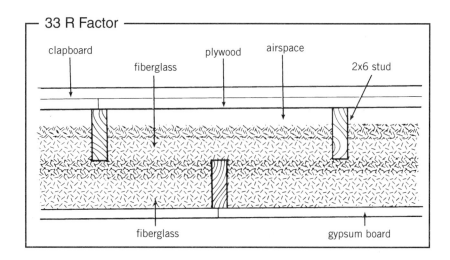

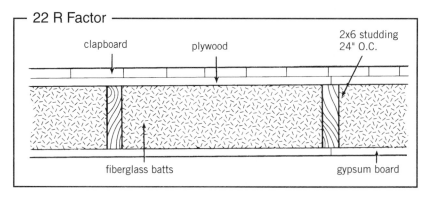

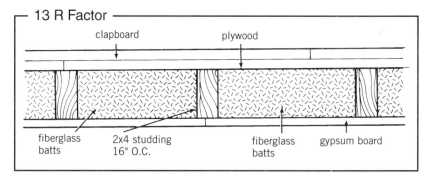

Three ways (and there are many more) of building outer walls, and their R factors. The top wall has an R factor of about 33, with the 2x6 studs not touching both the inner and outer layers. The middle example, with 2x6 studding, is about R-22. Above is the conventional method used until a few years ago, about R-13.

"or equal." This means that the actual item finally selected, such as a kitchen faucet, will be of an approximately equal dollar amount and of similar quality to the one in the specs.

Parts of Plans

Be sure your plans include the following pages, examples of which are shown here or in the appendix.

◊ **A plot plan.** If your plans are drawn locally, the designer can prepare this plan for you. Otherwise you or a surveyor will have to do it. It is merely a plat (map) of your lot with the position of the house drawn or blocked in. Its purpose is to ensure that a given house plan will fit on a given lot. If it doesn't fit on paper, it surely won't fit in reality. (Note: A plot plan is required during the building permit application.)

First, the lot is drawn, then all the setbacks required by zoning or other restrictions are sketched in, then the location of the house is indicated. The exact location may be slightly different when the house is physically staked out, but will be close to the indicated position on the plat — especially if it is a tight fit. If you have a lot consisting of several acres, the actual position of the house from that indicated on the plat could change by many feet.

The plot plan won't be drawn first if you have a set of plans but no lot. It's better to find a lot first and then make the plans to fit, though this is not an absolute rule. Some people love a certain set of plans or they want a particular house they have seen in a magazine. They would rather search for a lot to fit the plans. It can be done, although not always easily or inexpensively. (See page 29.)

◊ **Spec sheets.** In these lists of materials for your house, spell out as much detail as possible. Using an existing set (such as those found in the appendix on pages 126–136) is the easiest way to start. Make any necessary modifications.

◇ **Foundation plan.** This shows the overall dimensions of the house and the locations of all load-bearing requirements, such as piers, steel reinforcing rods, vents, basement slabs, and, if a basement house, all walls, windows, doors, and plumbing of that basement. (See page 140 in the appendix for a foundation plan.)

◇ **Floor plans.** You must have a plan for each floor that shows outside dimensions plus locations of windows and doors, plumbing fixtures, and large appliances. You may add the electrical drawings as well. These drawings show the locations of receptacles and switches. It is not necessary to have these shown on the floor plans. You can go through the house with your electrician after it is framed and locate receptacles, light fixtures, switches, cable outlets, and phone lines. However, when getting quotes or bids from subs, it is best to have the electrical, cable, and phone requirements included either in the floor plans or on a separate sheet.

The same is true for mechanical heating and cooling requirements. It is best to have these drawn out in advance, for bid purposes and to locate the furnace. If you plan to heat with wood, coal, gas, or oil, then planning the flue location is necessary, especially in a house without a basement or in a two-story house, unless your appliance is a direct-vent model.

◇ **Detail sheet.** This shows cabinet details, cross sections of the inside of the house if there is a section that may not be perfectly clear from the floor plan, and wall sections to show all the materials that make up a wall, such as brick, sheathing, studs, insulation, and inside wallboard or paneling. A detailed section of a foundation wall is also sketched in order to show proper construction technique, drainage, and waterproofing.

◇ **Outside elevations.** These sheets (usually two) show an outside view of how the house will look when finished. Usually all four sides are shown.

Plot plan

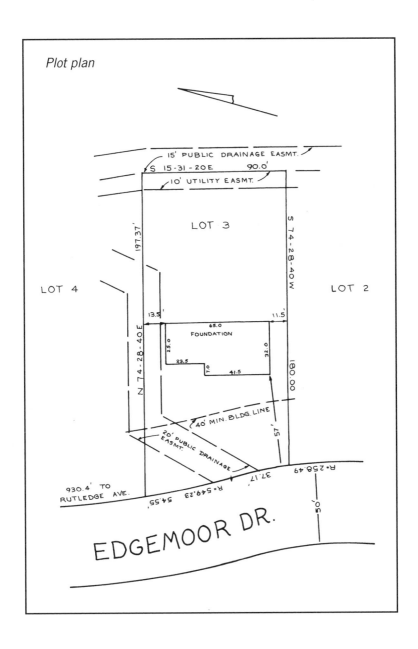

Cost, Appraisal, Decision!

Just a few steps remain in your decision-making process. Step 5, estimating the cost to build the house you have selected, is discussed thoroughly in the next chapter.

Step 6 involves getting an appraisal of the final value of your house. As I said in chapter 1, you can obtain a written appraisal of what your dream home will be worth after it is built on the land you have selected (or own). You will be getting such an appraisal ordered by your construction lender, but you can order one yourself for planning purposes as soon as you have selected land, plans, and basic specifications. Look in the Yellow Pages under Appraisers. It is not expensive to order an appraisal for yourself.

Now you can move on to Step 7: Decision time! Based on total cost of land, labor, and material, is this project worth it? With the appraisal in hand, you will be able to make your decision intelligently by reviewing the following factors:

1. Are you saving any money?
2. Is the house over, or under, priced for the neighborhood? (Under is better!)
3. Can you revise your cost estimate if there is less than 25 percent savings?

Once you decide to move ahead with your dream house and have obtained financing, you can start on the fun part of seeing the work actually begin.

Chapter 3

Cost Estimating

I USED TO SAY, "Estimating the cost of a new house is not an exact science." Now I believe that due to technology, it's *more* exact than some sciences. You can come very close to correctly estimating your total cost to build, and you can guide your construction project toward that estimated cost right up till the last cost expenditure, but you can only do it if *you* are the general contractor.

Why is cost estimating more accurate today? One word: spreadsheets. When I started in this business in 1970, I didn't know what a spreadsheet was. I found it difficult to manually list bids from subcontractors and suppliers, make adjustments for actual costs, and keep track of overages during construction. (There are rarely adjustments for underages!) However, over the years, I have seen the absolute necessity for cost control.

I try to stay abreast of technology as it relates to helping builders figure their cost estimates and track those costs during construction, and I have seen cost estimating become more accurate and easier to calculate. There are many spreadsheets available. Some are expensive or difficult to use, and most require expensive software, such as Microsoft Excel, which isn't always installed on home computers. So I designed a simple preformatted spreadsheet that is available at my Web site (see box on following page).

It is critical to do accurate cost estimating and even more important to control those costs as the house progresses. If you don't calculate an accurate budget *before* you start building and stick to that estimated budget, you probably won't save 25 percent of the cost of the project, and you may well end up not being able to finish or even keep the house. This is not good!

Be Your Own House Contractor
www.byoh.com

One of the main reasons for putting out a new edition of
this book is to include my new Web site, www.byoh.com,
where you'll find:

◇ Downloadable software to control building costs
◇ Up-to-date, average "cost to build" figures for the U.S.
◇ Current mortgage rates updated daily
◇ Newsletter with services, product reviews, new
  technology, and letters from other homebuilders
◇ Links to many useful sites with owner/builder
  construction lenders, house plans, site supervisors,
  and other valuable information

Remember: cost estimating is the most important step! If
you can't afford the house, now is the time to find out. My
Web site will help you accurately figure out the cost of a
new home in your area and give you up-to-date interest
rates to work with. The cost-estimating program comes
bundled with its own software and is very easy to use. It
(almost) makes estimating fun!

The Process of Estimating

Do you like to shop? If you don't, you may not save as much as someone who does. As I am fond of saying in my seminars, "Building a
house is the largest shopping expedition you will ever go on. You will
'shop till you drop' and you'll still be shopping as the moving van
pulls up to your new house. For some people, it's as if they died and
went to heaven. For others, well, . . . tough."

Putting together your bids and estimates requires that you shop
around. How do you shop for a whole house? Use the cost estimate
list on the opposite page as your shopping list.

Note: Get all bids and estimates in writing!

SUBCONTRACTOR, SUPPLIER	DESCRIPTION	ORIGINAL COST ESTIMATE	A ADJUSTED COST ESTIMATE	B TOTAL OF PREVIOUS PAYMENT	C CURRENT AMOUNTS TO BE PAID	D ESTIMATED COST TO COMPLETE
	Permits, fees, surveys					
	Utilities					
	Excavation					
	Foundation					
	Rough Lumber					
	Rough Labor					
	Windows & Exterior Doors					
	Roofing					
	Concrete Flatwork					
	Siding					
	Plumbing					
	Heating					
	Electrical					
	Insulation					
	Water (Well)					
	Sewer (Septic)					
	Fireplaces					
	Drywall					
	Cabinets					
	Interior Trim					
	Interior Trim Labor					
	Painting					
	Appliances					
	Light Fixtures					
	Floor Coverings					
	Driveway					
	Garage Door					
	Land					
	Misc.					
	Other					
TOTALS						

Note: Total of columns B+C+D should = Total of column A at all times

Let Your House Plans Do the Talking

Once you get your house plans, start shopping for bids and estimates from subcontractors and suppliers. The items you need fall into categories as seen on the cost estimate spreadsheet on page 33. These are suggested categories. You can alter them and add more as you see fit. Your objective is to get firm bids for all items.

Never accept a bid that is "by the hour." It doesn't work. Remember Murphy's Law #390: "If you want to see how long a job can take, pay them by the hour." You may pay a little more for a firm bid, but it's worth it for peace of mind. For example, if an excavator quotes $X per hour per man plus $X per hour per piece of equipment, insist on a firm total. If he won't offer one, move on to the next excavator on your list. See chapter 6 for finding good subcontractors.

The people and companies you will be contacting know how to give estimates based on plans and are used to being asked for bids. Don't worry, this is part of their job, whether you eventually hire them or not. I take my plans to a sub or supplier and say, "Give me a price on XXX. If you see anything else on the plans you can provide, give me a price on that, too."

One morning early in my building career, I met with an excavator at my lot to give him the plans and get a price for excavating. He gave me what I thought was a good price. He then said, "We also do septic systems, driveways, backfill, rough and final grading, and a few other things, and my brother does foundations, concrete slabs, and flat work." Wow, I struck gold! I hate to shop, so I got bids on four more parts of the project. See how it works?

Note that your suppliers and subcontractors will determine the exact (almost) number of items and square footage of materials needed based on your house plans. This is called a takeoff.

Every line item is also broken down as a suggested percentage of the total cost: see the Inspection Report and Disbursement Schedule on page 45, for a good idea of what to expect.

As you go through this process line by line, category by category, you may find even better prices for previous line items that you already have bids for. With the spreadsheet up and running, you'll be amazed how easy it is to watch the bottom line (the "total" line) shrink as you get the best prices.

You can make as many spreadsheets for subcategories within the line items as you need. For example: Foundation costs include several factors such as sand, fill dirt, steel reinforcing, and forms. Simply start a new spreadsheet, label it "Foundation," and change the list of costs by typing in what you need. The pre-typed ones can be overwritten.

Do a new one for each "Main" category as needed. It's fun!

"Whew!" you may be saying, "It's a lot of to work getting all those bids and estimates!" Well, I never said you didn't have to work to earn your 25 percent, but you still won't have to pick up a hammer!

When you finally see the bottom line of the first column approaching 75 percent of your budget, you should feel pretty good. As I have already mentioned, and will discuss further in chapter 4, your lender will order an appraisal that will tell you the market value of your completed house. Your estimated total cost of construction (that's the bottom line), excluding land, permits, and the 6 or 7 percent real estate commissions figured in by the appraiser, should not exceed 50 to 60 percent of that market value. If it's higher, get back to work shopping.

Building supply companies, home centers, carpet stores, appliance dealers, and so forth will give you the best price they can, as they know you are shopping around. If they don't, you'll know as you enter their bids on your spreadsheet and the bottom line grows too large or the prices don't match the suggested percent of total cost. (See the Inspection Report and Disbursement Schedule, page 45.)

Balancing Costs

As you pay for items during construction, enter the actual price paid under the "adjusted cost estimate" column on your spreadsheet. Scroll down to the "total" lines, compare "adjusted" to "original" and you'll know instantly if you are running over cost. If expenses run higher than estimated in some categories, you may have to trim costs in others.

Most of the heavy costs come at the beginning of construction for such items as excavation work, well and septic digging and installation, lumber, masonry, carpentry, plumbing, electrical work, windows

\d heating, ventilating, and air-conditioning (HVAC).
_.. items bought later can have a very significant effect on the total cost of the house. These include hardwood floors and other flooring, appliances, plumbing fixtures, wallpapering, millwork and interior trim, and carpeting. It is possible to cut costs on many of these items by choosing less expensive options or postponing installation of certain items (such as wallpaper) until after you've moved in.

Molding and trim details (such as a wainscot or cornice molding) can be added later if reducing immediate costs is important. Windows can vary considerably in price just because of a brand name, though the difference in quality is, in my opinion, negligible. Do you need a ten-cycle dishwasher, when a five-cycle or a two-cycle will do as well? So it is with each item. Compromise is the key to happy homebuilding!

> Some of the expensive items you may want in your house may add little, if any, to the appraised value (resale value). Examples are handmade tiles, exotic wood trims, and granite counters. Sorry, but that's a real estate fact of life.

Cost Breakdowns

Here is a more detailed description of the items and categories you will most likely encounter in building your home. If an accurate dollar amount estimate isn't available, use an educated guesstimate for now.

◆ **Permits, fees, surveys.** All costs necessary to get your project started can be included in this category and some guesstimates can be utilized. Your building inspection department can give you the cost of permits and fees, which vary by locale. Permits in some areas include water and sewer tap-in fees. Permits can be very expensive and time-consuming.

Your lender will require you to have a vacant land survey and another survey once the foundation is in, called a mortgage survey, which determines where the house actually is placed relative to property lines and setback requirements. (You should get one of these even if you are paying cash and don't have a lender.) You will also have to insure your building project against loss due to fire or other acts of God.

◊ **Installing Utilities (electric, gas, phone).** In some rural areas this item can be very costly, running into thousands of dollars. Check with your local utility companies.

◊ **Excavation.** This item depends on locale, soil conditions, terrain, and season of the year. Excavating for a basement will cost more than excavating for a slab or a crawl space.

◊ **Foundation.** This item will vary considerably based on factors such as slope of lot and the height of basement walls.

◊ **Rough Lumber.** See the list on page 80 for an idea of what goes into the "framing package," as it is called. This includes all materials except windows, doors, and roof shingles (although it can include those as well). These are most of the materials necessary to dry-in the house, which means putting up the walls, windows, exterior doors, and roof to make the interior waterproof. A good lumber company will put together this material list for you free of charge because it wants your business.

◊ **Rough Labor.** This is the labor required to bring the house to the dry-in stage. After this stage is completed, all other stages can commence, some simultaneously. The best way to contract for this job is by the square foot, with the square footage agreed to before you start. In determining the square footage, houses are measured from outside wall to outside wall, not from roof overhangs. Five people will arrive at five different square footage totals, using the same set of plans. Some will vary by 300 square feet or more. Sounds incredible, doesn't it? But I swear that it's true.

If the house is not easily divided into rectangles for simplifying square footage determination and you can't figure it out, have the designer do it for you. Ready-made plans generally come with the square footage broken down for you. Use those figures.

◇ **Windows and Exterior Doors.** This cost is simple to estimate since you have an exact count. I do not recommend any particular brand, but I do recommend that you visit a couple of building supply companies and compare. Most carry more than one brand. Locally made windows usually are less expensive than national brands and often carry comparable warranties. For special windows such as angular or bay windows, get exact quotes from the supplier. Generally there is no additional cost to install windows (except special or unusual ones), as that labor is included in your carpenter's framing charge. Be sure it is.

Exterior doors are as easy to estimate as windows, and all the same factors apply, including that the labor to install them should be included in the framing bill from your carpenter. Sliding glass or patio doors will be slightly more expensive and may require a separate installation charge, depending on your carpenter. Be sure that the installer caulks under the threshold thoroughly, using exterior grade silicone or polyurethane caulk.

◇ **Roofing.** This is measured and estimated by "squares." A square of roofing is the amount of roofing material required to cover 100 square feet (10 feet x 10 feet). I advise having either your supplier or your contractor do this estimate. It won't be exact, but may come a little closer than you will, though you might want to make your own calculations and see how they compare.

Asphalt-fiberglass shingles are priced according to the guarantee offered by the manufacturer. Fifteen-year shingles will be less expensive per square than 20- or 25-year shingles. It is probably best to go with a better-quality shingle, which will still be more economical than other types of roofing such as cedar shakes or steel panels. The cost of labor to install will depend upon the sub, the weight of the shingle (more durable shingles are usually heavier and therefore more expensive), and the pitch of the roof (the steeper the pitch, the higher the price for labor). The most common shingles for roofing are the 245-pound asphalt variety, and an average roof pitch is about 6 inches rise for each foot of horizontal travel (a 6/12 pitch).

◇ **Concrete flatwork (slabs), garage floors, basement floors.** This refers to smooth finish concrete work, not rough finish as in driveways, patios, and walks. It also involves the use of other materials such as Styrofoam, wire mesh, expansion joints, and polyethylene. Extra site preparation and gravel base installation can also figure in this expense. Your concrete subcontractors can explain this to you. The work is closely inspected by most building departments. Get a bid based on square footage of actual concrete area.

◇ **Siding.** Depending on the material used (which may range from brick to vinyl), you can get an accurate bid from a subcontractor that includes the cost of siding materials and any flashing required around windows and doors.

◇ **Plumbing.** This bid should include all fixtures such as toilets, sinks, and water heater. It will not include such items as a dishwasher, garbage disposal, washing machine, or other household appliances. If you supply the plumbing fixtures, your plumbing labor bids will be higher to offset the plumber's loss of profit on the fixtures. If a fixture you supply is defective, you will be responsible for taking care of warranty work, not the plumber. Keep that in mind.

◇ **Heating.** Use heating and air-conditioning systems recommended by your local utility company experts. Cost of installation should include proper ventilation for the bathrooms, the kitchen, and the clothes dryer.

◇ **Electrical.** In addition to electrical wiring costs, this bid should include all switches, receptacles, wires, panels, and breakers, wiring of all built-in appliances, cable and phone wiring, and compliance with codes. It does not include lighting fixtures.

◇ **Insulation.** Get a bid per local code for minimum insulation. To get the maximum insulation for number of dollars spent, consult your local utility company experts.

◇ **Water (Well).** For water tap-in fees, call your local municipality. For a well bid, call a well drilling firm familiar with the area and get a firm maximum bid as well as a price per linear foot drilled.

◇ **Sewer (Septic).** For a sewer tap-in fee, call your local municipality. For the price of a septic field, get a written bid from a local contractor who does septic system installation. **Note:** Alternative types of septic systems for clay soil or high water table are very expensive but usually doable.

◇ **Fireplaces.** Prefabricated fireplaces are less expensive; masonry is generally very expensive but may give you more options. Shop around for differences and determine what your budget can handle.

◇ **Drywall.** Bids should include labor and materials to hang the wallboard, tape joints, and finish with joint compound (two coats). If you supply the drywall, you won't save any money, but you will be responsible for scheduling delivery, so decide how much your time and stress is worth.

◇ **Cabinets.** Bids should include kitchen cabinets and bathroom vanity cabinets. Labor to install should be included in the carpenter's trim labor.

◇ **Interior Trim**. A bid from the lumber supplier should include all interior doors, moldings, closet shelves, and stairway trim. It should include additional subflooring (also called underlayment) for carpet.

◇ **Interior Trim Labor.** You should get a bid to install all the materials for both cabinets and interior trim.

◇ **Painting.** First-time house contractors (builders) often skimp on this category by planning to do it themselves. But it is still smart to get an estimate, even if just for the materials. If you are not planning on doing the labor yourself, get a written bid for both labor and materials. If you prefer to do your own labor, remember Murphy's Law #414: "Do-it-yourself labor takes twice as long and you get half the quality."

◇ **Appliances.** In the planning stage, you don't need to pick out the actual models, but you do need a ballpark idea of what you intend to buy. Use a dollar allowance in your estimate that you feel is adequate for the appliances you want.

◇ **Light Fixtures.** Though you probably won't have your actual fixtures selected, figure an amount that will be enough to cover the costs of all necessary fixtures. This can be an expensive category, so do some research by shopping around at lighting stores, "big box" home centers, and online.

◇ **Floor Coverings.** Estimate approximate costs for all floor coverings, taking into account varying amounts for wood, carpet, tile, or other coverings. As with other features, have in mind a price range for materials. At this early stage in planning, your tastes may change, or more likely, your budget may dictate a change in the final selection.

◇ **Driveway.** Depending on the material used, get a bid based on the square footage of the area to be covered. Make sure you agree on the square footage.

◇ **Garage Door.** If you can still afford one, get a price from your local lumber company — with or without operators, installed or uninstalled.

◇ **Other.** Anything not included above, such as fences, sidewalks, decks, swimming pools, saunas, and so on.

Chapter 4

Financing

THE MOST COMMON WAY TO FINANCE the building of a home is to obtain a construction loan, which is a short-term line of credit for paying construction costs. There are two kinds of construction loans.

The one-time close construction loan: As the name implies, this type of loan entails one closing (signing papers) for both the short-term construction line of credit and the "permanent" end financing (mortgage).

One closing means one set of closing costs, which is the main advantage for this type of loan. Since the lender's policies on these types of loans vary so much and change so fast, you will need to talk directly to a lender when you are ready to shop for your loan.

The main disadvantage to this loan is that the dollar amount of both the construction line of credit and the "permanent" end financing (mortgage) cannot be increased. If the house ends up costing more than you planned, and your loan isn't large enough, you will end up taking out another loan and paying for a second closing anyway and you may find yourself trapped and at the mercy of the lender.

Note: These loans are not always available to owner/builders as lenders feel that an owner/builder cannot estimate costs accurately and is not a good risk.

The two-time close construction loan: This type of loan involves two separate loans with two sets of closing costs, which offers you more flexibility. If you do decide to increase the amount of your construction loan (and if you qualify), it's no big deal. You will pay off the construction loan with a refinance mortgage. Refinance mortgages are very competitive, and you might get a better deal by being able to "shop" for the loan that you'll be paying on for years.

I prefer the two-time construction loan. There is less pressure on me as the builder, and it can actually be less expensive in the long run.

How a Construction Loan Works

A construction loan is like a credit card with a low interest rate and a high credit limit. If you were able to secure a mortgage for the construction of your new house, and the lender were to give you a check for that amount before you started, both you and the lender would be in trouble. The lender, because it just loaned a large sum of money on a house that doesn't yet exist; you, because you'd have to start paying interest and principal on that loan before you even broke ground for your new house. Considering the interest expense, this could mean you would have to pay a large amount of interest each month on a nonexistent house, along with your regular monthly fixed expenses for items such as food and rent.

Enter the construction loan. The money you need to pay for construction costs, even the land in some cases, is disbursed to you as the building progresses. You pay only interest, no principal, on the total amount lent to you for any given month during construction. For example, if the house is 20 percent complete after the first month, the lender, after a physical inspection of the progress of construction, will advance 20 percent of the total loan to you to pay your bills. The Inspection Report and Disbursement Schedule on page 45 shows how this percentage is determined. If your total mortgage is $X, then you will receive 20 percent of $X.

The house is now 20 percent complete, you are one month along in progress, and you have paid no interest yet. And you won't have to until next month, when you will start to pay the interest only on the money you received.

The application process with a good lender should be quick and painless. To find the best deal, talk to several lenders that offer both types of loans.

Construction loan interest is a "cost to build" item of any new home, whether you build it or I build it for you. It's not even part of the builder's overhead. It's a cost like lumber and nails. Treat it as such and plan for it, along with other financing charges, and it won't bother you. Think of all the money you are saving by being your own general contractor. Be aware, however, that many lenders will not permit loan interest to be paid out of the loan itself.

There is a considerable time period during construction in which the house is progressing toward completion, but you haven't yet received bills for materials. It should take approximately six months to complete the house, and because of the billing time lag from suppliers and the loan interest being one month or more in arrears, during that six months you will have had an average of only 40 percent of the total money available disbursed to you. This is an approximate figure, obviously, as is the amount of time needed to complete your house. I'm merely pointing out that you won't pay interest on the full amount of your mortgage until the house is complete. At that time the lender will complete the disbursement of the construction money and convert the loan to a permanent mortgage for 15 to 30 years, or you can refinance the construction loan with the "permanent" end financing (mortgage) that you were pre-approved for. Only then will you be making your full payment of principal and interest. But by then you will have moved in to your dream house. If that sounds simple, it is only because it is.

Two Homes?

By the way, even in a sluggish market, all houses do sell sooner or later, but if you already have a home and you fear it won't sell before your new one is ready, go ahead and sell it first and live in an apartment while building your new home. Also, you may not qualify for a construction loan with an existing house payment. Talk to a lender.

Financing charges vary with lenders but will be estimated in advance for you. They will probably total about 2 to 3 percent of the mortgage amount. Charges may have to be paid when you close your construction loan, or they may be deducted from the total construction loan amount.

INSPECTION REPORT AND DISBURSEMENT SCHEDULE

Borrower _____

Contractor _____

Job address _____

City _____

	% OF TOTAL	MONTHLY INSPECTIONS				
		1	2	3	4	5
1. Excavation	2					
2. Foundation & slab work	8					
3. Floor	4					
4. Wall framing	6					
5. Roof framing & sheeting	6					
6. Wall sheathing	1					
7. Roofing	2					
8. Windows set	10					
9. Exterior doors	2					
10. Siding/brick	7					
11. Exterior trim	1					
12. Well/water tap	2					
13. Septic/sewer tap	2					
14. Plumbing roughed	5					
15. Wiring roughed	2					
16. Heating roughed	2					
17. Heat plant	2					
18. Insulation	2					
19. Chimney/flue	2					
20. Drywall	6					
21. Bath tile	1					
22. Plumbing fixtures	4					
23. Interior trim	3					
24. Cabinets	3					
25. Exterior painting	2					
26. Interior painting	2					
27. Appliances	2					
28. Light fixtures	1					
29. Carpet/floor finish	4					
30. Drives and walks	3					
31. Landscaping	1					
TOTAL	100%					

Inspector _____

Comments: _____

Cash Needed

You will need some extra cash. Call it interim (short-term) working capital. Some expenses will occur before your lender makes the first disbursement of construction money, and you will need a few thousand dollars to bridge that gap for such things as paying subs, fees, and permits. If you have cash available, fine. If not, you may want to borrow from a commercial bank or take a second mortgage on your existing home or get a bridge loan.

Excluding the cost of the land and the 2 to 3 percent in financing charges I discussed, you shouldn't need more than an additional 5 percent of the cost of the house for interim operating expenses. This money is needed only until the house is finished. When the lender disburses all the money in the form of the permanent loan, the need for your operating cash will disappear and you can put it back where you got it.

Lenders will have different opinions on many of your particular financing needs, and you need to shop around and spend some time talking to them. They will help you tremendously in putting your financial package together. In this aspect, they are like subcontractors. Each of them will also explain their policies on construction draws, inspections, construction loan interest payments, points, credit reports, qualifications, and interest rates.

Qualifying for a Loan

The requirements for qualifying for a loan change periodically and differ for each lender. Any mortgage lender can give you a preliminary estimate of your qualifications over the phone. In general, the better your credit, the more money you put down (increasing your equity), and the lower your total debts are in relationship to your total income (called debts to income or DTI), the more a lender will be willing to lend, and often at a lower interest rate. You should only be applying for a loan that is approximately 75 to 80 percent of the "subject to completion" appraised value of your project. This is considered a low loan-to-value loan (LTV). I mentioned earlier that your lender would be obtaining such an appraisal as part of the loan application process.

Consumer Reports 2020 Car Giveaway

Ticket issued to:
W. Anderson

Suggested donation:
9 Tickets at $20

Ticket valid for 1 (one) entry into Consumer Reports 2020 Car Giveaway. Deadline: 7/1/20.

TICKET NUMBER
UB641951

KEEP TICKETS · RETURN ORDER FORM AT LEFT

UB641951

Consumer Reports 2020 Car Giveaway

Ticket issued to:
W. Anderson

Suggested donation:
9 Tickets at $20

Ticket valid for 1 (one) entry into Consumer Reports 2020 Car Giveaway. Deadline: 7/1/20.

TICKET NUMBER
UB641952

KEEP TICKETS · RETURN ORDER FORM AT LEFT

UB641952

Consumer Reports 2020 Car Giveaway

Ticket issued to:
W. Anderson

Suggested donation:
9 Tickets at $20

Ticket valid for 1 (one) entry into Consumer Reports 2020 Car Giveaway. Deadline: 7/1/20.

TICKET NUMBER
UB641953

KEEP TICKETS · RETURN ORDER FORM AT LEFT

UB641953

$5,000

(1) **First Prize**	$5,000
(2) **Second Prizes**	$1,000 each
(5) **Third Prizes**	$500 each
(10) **Fourth Prizes**	$250 each
(100) **Fifth Prizes**	valued at $24 each
Plus, (1) **Early Bird Prize of $1,500**	

The end loan, or permanent mortgage application, is always part of the construction loan application process, regardless of whether or not you obtain a one-time close or two-time close construction loan. The number of end loan programs available today is staggering! You can find everything from 30- and 15-year fixed (very traditional) to adjustable rates tied to various markets to "interest only" mortgages. There are hundreds of programs where 25 years ago, there were half a dozen. This is good for you as a borrower!

Paying for the Land

Paying for your building site used to be a nightmare. If you didn't have the cash to pay for the land up front, you had a problem, as the land usually has to be paid off in order for construction financing to be approved. Lenders typically did not include the land in the construction financing, but now some of them are willing to do so.

Lenders typically would loan you 80 percent (or less) of your acquisition costs alone, which includes land, labor, and materials. These days, some companies will lend you 80 percent (or more) of what is called the "subject to completion" appraised value (STC appraisal). The STC is based on a professional appraiser's opinion of what the true market value of your house and land will be when construction is completed.

To see this more clearly, let's say that the STC appraised value is $100,000. Residential construction usually fits into the following formula: 25 percent each for land cost, labor, material, and builder's profit and overhead. If you can only borrow 80 percent of the first three items, you have $60,000 (80 percent of $75,000). This isn't enough to cover your costs. If, however, you can borrow 80 percent of the STC appraised value, you can borrow $80,000, which should cover all your costs, including land payoff. This is the ideal situation. Try to come as close to it as you can or need to. Do some research, as lenders' policies are like snowflakes: No two are alike and they disappear fast.

STANDARD MORTGAGE APPLICATION FORM

Uniform Residential Loan Application

This application is designed to be completed by the applicant(s) with the lender's assistance. Applicants should complete this form as "Borrower" or "Co-Borrower", as applicable. Co-Borrower information must also be provided (and the appropriate box checked) when ☐ the income or assets of a person other than the "Borrower" (including the Borrower's spouse) will be used as a basis for loan qualification or ☐ the income or assets of the Borrower's spouse will not be used as a basis for loan qualification, but his or her liabilities must be considered because the Borrower resides in a community property state, the security property is located in a community property state, or the Borrower is relying on other property located in a community property state as a basis for repayment of the loan.

I. TYPE OF MORTGAGE AND TERMS OF LOAN

Mortgage Applied for:	☐ VA ☐ FHA	☐ Conventional ☐ FmHA	☐ Other:	Agency Case Number	Lender Case No.

Amount	Interest Rate	No. of Months	Amortization Type:	☐ Fixed Rate ☐ GPM	☐ Other (explain): ☐ ARM (type):
$	%				

II. PROPERTY INFORMATION AND PURPOSE OF LOAN

Subject Property Address (street, city, state, & zip code)	No. of Units

Legal Description of Subject Property (attach description if necessary)	Year Built

Purpose of Loan ☐ Purchase ☐ Construction ☐ Other (explain): ☐ Refinance ☐ Construction-Permanent

Property will be: ☐ Primary Residence ☐ Secondary Residence ☐ Investment

Complete this line if construction or construction-permanent loan.

Year Lot Acquired	Original Cost	Amount Existing Liens	(a) Present Value of Lot	(b) Cost of Improvements	Total (a + b)
	$	$	$	$	$

Complete this line if this is a refinance loan.

Year Acquired	Original Cost	Amount Existing Liens	Purpose of Refinance	Describe Improvements ☐ made ☐ to be made
	$	$		Cost: $

Title will be held in what Name(s)	Manner in which Title will be held	Estate will be held in: ☐ Fee Simple

Source of Down Payment, Settlement Charges and/or Subordinate Financing (explain)	☐ Leasehold (show expiration date)

III. BORROWER INFORMATION

Borrower	Co-Borrower
Borrower's Name (include Jr. or Sr. if applicable)	Co-Borrower's Name (include Jr. or Sr. if applicable)

Social Security Number	Home Phone (incl. area code)	Age	Yrs. School	Social Security Number	Home Phone (incl. area code)	Age	Yrs. School

☐ Married ☐ Separated ☐ Unmarried (include single, divorced, widowed)	Dependents (not listed by Co-Borrower) no. ages	☐ Married ☐ Separated ☐ Unmarried (include single, divorced, widowed)	Dependents (not listed by Borrower) no. ages

Present Address (street, city, state, zip code) ☐ Own ☐ Rent ___ No. Yrs.	Present Address (street, city, state, zip code) ☐ Own ☐ Rent ___ No. Yrs.

If residing at present address for less than two years, complete the following:

Former Address (street, city, state, zip code) ☐ Own ☐ Rent ___ No. Yrs.	Former Address (street, city, state, zip code) ☐ Own ☐ Rent ___ No. Yrs.

Former Address (street, city, state, zip code) ☐ Own ☐ Rent ___ No. Yrs.	Former Address (street, city, state, zip code) ☐ Own ☐ Rent ___ No. Yrs.

IV. EMPLOYMENT INFORMATION

Borrower	Co-Borrower

Name & Address of Employer ☐ Self Employed	Yrs. on this job	Name & Address of Employer ☐ Self Employed	Yrs. on this job
	Yrs. employed in this line of work/profession		Yrs. employed in this line of work/profession

Position/Title/Type of Business	Business Phone (incl. area code)	Position/Title/Type of Business	Business Phone (incl. area code)

If employed in current position for less than two years or if currently employed in more than one position, complete the following:

Name & Address of Employer ☐ Self Employed	Dates (from - to)	Name & Address of Employer ☐ Self Employed	Dates (from - to)
	Monthly Income $		Monthly Income $

Position/Title/Type of Business	Business Phone (incl. area code)	Position/Title/Type of Business	Business Phone (incl. area code)

Name & Address of Employer ☐ Self Employed	Dates (from - to)	Name & Address of Employer ☐ Self Employed	Dates (from - to)
	Monthly Income $		Monthly Income $

Position/Title/Type of Business	Business Phone (incl. area code)	Position/Title/Type of Business	Business Phone (incl. area code)

V. MONTHLY INCOME AND COMBINED HOUSING EXPENSE INFORMATION

Gross Monthly Income	Borrower	Co-Borrower	Total	Combined Monthly Housing Expense	Present	Proposed
Base Empl. Income*	$	$	$	Rent	$	
Overtime				First Mortgage (P&I)		$
Bonuses				Other Financing (P&I)		
Commissions				Hazard Insurance		
Dividends/Interest				Real Estate Taxes		
Net Rental Income				Mortgage Insurance		
Other (before completing, see the notice in "describe other income," below)				Homeowner Assn. Dues		
				Other:		
Total	$	$	$	Total	$	$

* Self Employed Borrower(s) may be required to provide additional documentation such as tax returns and financial statements.

Describe Other Income *Notice:* **Alimony, child support, or separate maintenance income need not be revealed if the Borrower (B) or Co-Borrower (C) does not choose to have it considered for repaying this loan.**

B/C		Monthly Amount
		$

VI. ASSETS AND LIABILITIES

This Statement and any applicable supporting schedules may be completed jointly by both married and unmarried Co-Borrowers if their assets and liabilities are sufficiently joined so that the Statement can be meaningfully and fairly presented on a combined basis; otherwise separate Statements and Schedules are required. If the Co-Borrower section was completed about a spouse, this Statement and supporting schedules must be completed about that spouse also.

Completed ☐ Jointly ☐ Not Jointly

ASSETS Description	Cash or Market Value	Liabilities and Pledged Assets. List the creditor's name, address and account number for all outstanding debts, including automobile loans, revolving charge accounts, real estate loans, alimony, child support, stock pledges, etc. Use continuation sheet, if necessary. Indicate by (*) those liabilities which will be satisfied upon sale of real estate owned or upon refinancing of the subject property.	Monthly Payt. & Mos. Left to Pay	Unpaid Balance
Cash deposit toward purchase held by:				
		LIABILITIES		
		Name and address of Company	$ Payt./Mos.	$
List checking and savings accounts below				
Name and address of Bank, S&L, or Credit Union				
		Acct. no.		
		Name and address of Company	$ Payt./Mos.	$
Acct. no.	$			
Name and address of Bank, S&L, or Credit Union				
		Acct. no.		
		Name and address of Company	$ Payt./Mos.	$
Acct. no.	$			
Name and address of Bank, S&L, or Credit Union				
		Acct. no.		
		Name and address of Company	$ Payt./Mos.	$
Acct. no.	$			
Name and address of Bank, S&L, or Credit Union				
		Acct. no.		
		Name and address of Company	$ Payt./Mos.	$
Acct. no.	$			
Stocks & Bonds (Company name/number & description)	$			
		Acct. no.		
		Name and address of Company	$ Payt./Mos.	$
Life insurance net cash value	$			
Face amount: $				
Subtotal Liquid Assets	$			
Real estate owned (enter market value from schedule of real estate owned)	$	Acct. no.		
Vested interest in retirement fund	$	Name and address of Company	$ Payt./Mos.	$
Net worth of business(es) owned (attach financial statement)	$			
Automobiles owned (make and year)	$			
		Acct. no.		
		Alimony/Child Support/Separate Maintenance Payments Owed to:	$	
Other Assets (itemize)	$	Job Related Expense (child care, union dues, etc.)	$	
		Total Monthly Payments	$	
Total Assets a.	$	**Net Worth** ➡ $ (a minus b)	**Total Liabilities b.**	$

(continued on next page)

Standard Mortgage Application Form (continued)

VI. ASSETS AND LIABILITIES (cont.)

Schedule of Real Estate Owned (If additional properties are owned, use continuation sheet.)

Property Address (enter S if sold, PS if pending sale or R if rental being held for income)	Type of Property	Present Market Value	Amount of Mortgages & Liens	Gross Rental Income	Mortgage Payments	Insurance, Maintenance, Taxes & Misc.	Net Rental Income
		$	$	$	$	$	$
Totals		$	$	$	$	$	$

List any additional names under which credit has previously been received and indicate appropriate creditor name(s) and account number(s):

Alternate Name	Creditor Name	Account Number

VII. DETAILS OF TRANSACTION

a. Purchase price	$
b. Alterations, improvements, repairs	
c. Land (if acquired separately)	
d. Refinance (incl. debts to be paid off)	
e. Estimated prepaid items	
f. Estimated closing costs	
g. PMI, MIP, Funding Fee	
h. Discount (if Borrower will pay)	
i. Total costs (add items a through h)	
j. Subordinate financing	
k. Borrower's closing costs paid by Seller	
l. Other Credits (explain)	
m. Loan amount (exclude PMI, MIP, Funding Fee financed)	
n. PMI, MIP, Funding Fee financed	
o. Loan amount (add m & n)	
p. Cash from/to Borrower (subtract j, k, l & o from i)	

VIII. DECLARATIONS

If you answer "yes" to any questions a through i, please use continuation sheet for explanation.

Borrower: Yes No — Co-Borrower: Yes No

a. Are there any outstanding judgments against you?

b. Have you been declared bankrupt within the past 7 years?

c. Have you had property foreclosed upon or given title or deed in lieu thereof in the last 7 years?

d. Are you a party to a law suit?

e. Have you directly or indirectly been obligated on any loan which resulted in foreclosure, transfer of title in lieu of foreclosure, or judgment? (This would include such loans as home mortgage loans, SBA loans, home improvement loans, educational loans, manufactured (mobile) home loans, any mortgage, financial obligation, bond, or loan guarantee. If "Yes," provide details, including date, name and address of Lender, FHA or VA case number, if any, and reasons for the action.)

f. Are you presently delinquent or in default on any Federal debt or any other loan, mortgage, financial obligation, bond, or loan guarantee? If "Yes," give details as described in the preceding question.

g. Are you obligated to pay alimony, child support, or separate maintenance?

h. Is any part of the down payment borrowed?

i. Are you a co-maker or endorser on a note?

j. Are you a U.S. citizen?

k. Are you a permanent resident alien?

l. Do you intend to occupy the property as your primary residence? If "Yes," complete question m below.

m. Have you had an ownership interest in a property in the last three years?

(1) What type of property did you own—principal residence (PR), second home (SH), or investment property (IP)?

(2) How did you hold title to the home—solely by yourself (S), jointly with your spouse (SP), or jointly with another person (O)?

IX. ACKNOWLEDGMENT AND AGREEMENT

The undersigned specifically acknowledge(s) and agree(s) that: (1) the loan requested by this application will be secured by a mortgage or deed of trust on the property described herein; (2) the property will not be used for any illegal or prohibited purpose or use; (3) all statements made in this application are made for the purpose of obtaining the loan indicated herein; (4) occupation of the property will be as indicated above; (5) verification or reverification of any information contained in the application may be made at any time by the Lender, its agents, successors and assigns, either directly or through a credit reporting agency, from any source named in this application, and the original copy of this application will be retained by the Lender, even if the loan is not approved; (6) the Lender, its agents, successors and assigns will rely on the information contained in the application and I/we have a continuing obligation to amend and/or supplement the information provided in this application if any of the material facts which I/we have represented herein should change prior to closing; (7) in the event my/our payments on the loan indicated in this application become delinquent, the Lender, its agents, successors and assigns, may, in addition to all their other rights and remedies, report my/our name(s) and account information to a credit reporting agency; (8) ownership of the loan may be transferred to successor or assign of the Lender without notice to me and/or the administration of the loan account may be transferred to an agent, successor or assign of the Lender with prior notice to me; (9) the Lender, its agents, successors and assigns make no representations or warranties, express or implied, to the Borrower(s) regarding the property, the condition of the property, or the value of the property.

Certification: I/We certify that the information provided in this application is true and correct as of the date set forth opposite my/our signature(s) on this application and acknowledge my/our understanding that any intentional or negligent misrepresentation(s) of the information contained in this application may result in civil liability and/or criminal penalties including, but not limited to, fine or imprisonment or both under the provisions of Title 18, United States Code, Section 1001, et seq. and liability for monetary damages to the Lender, its agents, successors and assigns, insurers and any other person who may suffer any loss due to reliance upon any misrepresentation which I/we have made on this application.

Borrower's Signature	Date	Co-Borrower's Signature	Date
X		X	

X. INFORMATION FOR GOVERNMENT MONITORING PURPOSES

The following information is requested by the Federal Government for certain types of loans related to a dwelling, in order to monitor the Lender's compliance with equal credit opportunity, fair housing and home mortgage disclosure laws. You are not required to furnish this information, but are encouraged to do so. The law provides that a Lender may neither discriminate on the basis of this information, nor on whether you choose to furnish it. However, if you choose not to furnish it, under Federal regulations this Lender is required to note race and sex on the basis of visual observation or surname. If you do not wish to furnish the information, please check the box below. (Lender must review the above material to assure that the disclosures satisfy all requirements to which the Lender is subject under applicable state law for the particular type of loan applied for.)

BORROWER ☐ I do not wish to furnish this information

Race/National Origin: ☐ American Indian or Alaskan Native ☐ Asian or Pacific Islander ☐ Black, not of Hispanic origin ☐ Hispanic ☐ White, not of Hispanic origin ☐ Other (specify) _____

Sex: ☐ Female ☐ Male

CO-BORROWER ☐ I do not wish to furnish this information

Race/National Origin: ☐ American Indian or Alaskan Native ☐ Asian or Pacific Islander ☐ Black, not of Hispanic origin ☐ Hispanic ☐ White, not of Hispanic origin ☐ Other (specify) _____

Sex: ☐ Female ☐ Male

To be Completed by Interviewer	Interviewer's Name (print or type)	Name and Address of Interviewer's Employer
This application was taken by: ☐ face-to-face interview ☐ by mail ☐ by telephone	Interviewer's Signature — Date	
	Interviewer's Phone Number (incl. area code)	

Chapter 5

Further Preparations

YOU ARE MAKING GREAT PROGRESS. You have purchased a lot, selected the house plans, estimated your costs, and arranged your construction loan and permanent mortgage. Several details still need your attention before you break ground.

By now you should have found out from your building inspection department the procedure and cost for obtaining your building permit. The sample building permit on pages 54–55 gives an example of information you will need and the inspections that will be required. Ask your local department about regulations in your area. Good subcontractors will take care of the inspections for you.

Inspections Required

Inspection requirements vary from place to place, but as an example, the county I live in requires the following inspections.

◇ Temporary electrical, or saw service, to ensure proper grounding.

◇ Footing, before pouring concrete, to make sure you have reached solid load-bearing ground.

◇ Slab, before concrete is poured, to determine whether it is properly insulated. Any plumbing in the slab is also inspected.

◇ Electrical, plumbing, heating, and air-conditioning rough-in, to ensure that in-the-wall installations are correct before they are covered up.

◇ Framing rough-in, to ensure structural integrity, especially after electrical, plumbing, heating, and air-conditioning installations. Workers have been known to cut too deeply into a joist or stud and weaken it.

◇ Insulation, to ensure compliance with local standards.

◇ Final electrical, plumbing, HVAC, and building to ensure systems work properly, comply with codes, and are safe. All final inspections must be completed before residents can get more than temporary electrical service. A certificate of occupancy (C of O) will be issued when all inspections are complete.

Final Details

If you haven't already done so, now is the time to select materials such as brick, shingles, windows, doors, roofing, trim, plumbing, fixtures, built-in appliances, light fixtures, flooring, hardware, and wallpaper. As you visit building supply houses to see their samples, you can also take care of a couple of other important tasks.

◇ Open a line of credit with them. This is quite easy, even for an inexperienced builder. See chapter 7 for a list of the types of suppliers you'll need.

◇ Ask them to recommend reliable subcontractors. Smaller building supply companies are better equipped to furnish this information because the management is more directly involved with customers. Remember, the key sub you are looking for is your carpenter, and most of the good ones patronize building supply houses. Many make building supply houses their source of referrals. This is especially true in rural resort areas. (Don't forget, you can also use this book as a guide to building a second home in a resort area.) You may find other subs here as well. Refer to the next chapter for more about subs.

◇ Now is the time to contact your gas, electric, and water departments for hookup procedures. Every locale has different procedures. You may also want to order a pre-wire from your phone, cable television, and broadband Internet company at this time, as these companies sometimes require three to five weeks' notice. Some don't charge for this, some do. Ask first.

◇ You will also want to shop around for a builder's risk or fire policy from various insurance agents. In most states, agents offer fairly equal policies, since they're governed by state laws. But be sure to check. The policy should take effect when materials arrive on the job, but your lender may want it to be in force before closing the construction loan. The lender will be the payee in the policy since it is the company's money. When the house is complete, and you move in, the fire policy can be converted to homeowner's insurance, often at a fairly good savings over a new homeowner's policy.

Doing Your Own Labor

If you plan to do any of your own labor, I have only one thing to say: If you aren't an expert at the particular trade you plan to do yourself, forget it. This is especially true in painting. People think it is easy, but unfortunately it comes at a time when the house is quite far along, and construction disbursements will be approaching their maximum amount. You need painters who can get in and get out, do the work fast, and do a respectable job. If you seek perfection in painting, wait until you are living in the house and do your own touch-up painting.

If you hire painters by the hour, plan to stay with them the whole time or plan on it taking twice as long. You won't get a better job for the extra time. Hire them by a contract amount, such as X dollars per square foot based on the amount of heated area.

When it comes to doing your own work, remember Murphy's Law #555 — "Nothing is as easy as it looks."

BUILDING PERMIT
DEPARTMENT OF BUILDINGS AND SAFETY ENGINEERING

The Bureau of Licenses and Permits
hereby grants permission to

Permit No. ———————————

Date ———————— 20———

_____
name   address   contractor's state license

_____
to   ( ) story   number of apartments

_____
on the   side of (building no.) street or avenue   zoning district

_____
lot no. and subdivision   size

Building is to be ——— ft. wide by ———ft. long by ———ft. in height
and shall conform in construction to type

Use group ———————————— Basement walls or foundation ——————

Remarks: ————————————————————————————————

————————————————————————————————

Approved by ——————————————————————————————

cubic feet ——— estimated cost $——— regular fees $——— zoning fee $———

Owner ———————————————   Bureau of Licenses and Permits

Address ————————————   by ———————————————

This permit conveys no right to occupy any street, alley, or sidewalk or any part
thereof, either temporarily or permanently. Encroachments on public property,
not specifically permitted under the building code, must be approved by the
common council. Street or alley grades as well as depth and location of public
sewers may be obtained from the Department of Public Works — City
Engineer's Office. The issuance of this permit does not release the application
from the conditions of any applicable subdivision restrictions. Minimum of
three call inspections required for all construction work:

1. Foundations or footings.
2. Prior to covering structural members (ready to lath).
3. Final inspection for compliance prior to obtaining certificate of occupancy.

Approved plans must be retained at the job site and this card must be posted
until final inspection has been made. Where a certificate of occupancy is
required, such building shall not be occupied until final inspection has been
made and certificate obtained.

Separate permits required for electrical and plumbing installations
POST THIS CARD

Building Inspection Approvals

1

Drain tile and foundation _____

Date _____

Inspector _____

2

Superstructure _____

(prior to lath and plaster)

Date _____

Inspector _____

3

Final inspection _____

Date _____

Inspector _____

Work shall not proceed until each bureau has approved the various stages of construction.

Electrical Inspection Approvals

1

Roughing in _____

Date _____

Inspector _____

2

Final inspection _____

Inspector _____

Safety Engineering Approval

3

Approved _____

Date _____

Inspector_____

Plumbing Inspection Approvals

1

Building Sewer _____

(A) Sanitary _____

Date _____

Inspector_____

(B) Storm _____

Date _____

Inspector _____

2

Crock to iron _____

Date _____

Inspector_____

3

Underground storm drains_____

Date_____

Inspector _____

4

Rough plumbing _____

Date _____

Inspector _____

5

Water piping final inspection _____

Date _____

Inspector_____

Inspections indicated on this card can be arranged for by telephone or written notification.

Chapter 6

Subcontractors

A SUBCONTRATOR (SUB) IS AN INDIVIDUAL or a firm that contracts to perform part or all of another contract. In your case you are technically the builder or general contractor, and you will build your house by subcontracting with others for specific jobs. You will pay for the project by setting a predetermined contract amount with each sub. This is important. You have no hourly wage employees working for you, which means you will avoid all of the governmental red tape and taxes concerning employees. Your subcontractors are not considered to be employees.

Your Subcontractors and Professionals

Here is a list of the subcontractors and professional people you probably will be contracting with, listed generally in the order in which you will need them:

- ◇ Real estate broker for land search
- ◇ Loan officer at bank, credit union, or mortgage lender
- ◇ House designer or architect
- ◇ Framing carpenter (your key sub, to be lined up early)
- ◇ Surveyor
- ◇ Grading and excavation contractor
- ◇ Soil treatment firm
- ◇ Footing contractor

◇ Brick masonry contractor to build a block foundation

◇ Concrete contractor to pour concrete walls and the slab or concrete floors, as well as drives, walks, and approaches

◇ Waterproofing contractor

◇ Electrical contractor

◇ Plumbing contractor (also well and septic system, if needed)

◇ Heating, ventilation, and air-conditioning (HVAC) contractor

◇ Roofing contractor

◇ Insulation contractor

◇ Drywall contractor

◇ Painting contractor

◇ Finish carpenter (Installs kitchen and other built-in cabinets and trim around doors and windows. This work is often, but not always, done by main carpentry sub.)

◇ Flooring, carpet, and countertop contractor

◇ Tile contractor

◇ Cleaning crew contractor

◇ Landscape contractor

◇ Licensed fee appraiser

Finding Your Subs

As mentioned earlier, a lumber supply store is the best place to start for finding your carpenter subcontractor. Your carpentry sub will be able to recommend almost everyone else, as he is on the job more than the others and knows most of the other subs involved in building a house.

"A good sub is a working sub," especially during a recession or other downturn in housing starts. This is not always true, but it is a pretty safe bet. The really good ones are sought after and always busy because they do good work and are reliable.

If you can't find a sub through a supplier or your carpenter, the next best place to look is on a job site. Find a house under construction. Stop and ask around. You can get names, prices, and references. This takes only a few minutes. It is done all the time, and the general contractor shouldn't mind. He probably won't even be there. Besides, he does the very same thing.

Often the boss or owner of a subcontracting firm is on the job. Get his number and arrange a meeting. Sometimes there are signs at the job site advertising different subs.

Carpentry Labor Contract

To (your name) _____ Subcontractor _____

(address)_____

_____

Date _____ Job address _____

Owner _____

Area _____ Heated _____ sq. ft.

Unheated _____ sq. ft.

Decks _____ sq. ft.

Charges

Framing   @_____ sq. ft x _____ sq. ft. = $ _____

Boxing and siding @_____ sq. ft x _____ sq. ft. = $ _____

Interior trim   @_____ sq. ft x _____ sq. ft. = $ _____

Decks   @_____ sq. ft x _____ sq. ft. = $ _____

Setting fireplace $ _____

Setting cabinets $ _____

Paneling $ _____

Misc. $ _____

Total charges $ _____

Signed (your name) _____ Date _____

Signed (subcontractor) _____ Date _____

Only certain trades of subs are listed in the Yellow Pages. Most independent carpenters are not listed, but you should be able to find heating and air-conditioning companies, plumbers, electricians, roofers, waterproofing companies, lumber dealers, appliance manufacturers, and a few others.

I once hired someone who was probably one of the best drywall subcontractors in the Southeast, but he was not listed. He didn't live in the same city; he lived out in the country. He was so good that he stayed backlogged two or three weeks even during recessions. His name was given to me by a lumber dealer.

Each subcontractor should carry insurance on his or her employees and should provide you with a certificate of insurance. (See the example in the appendix on page 137–138.)

Since this is your first experience and you won't be familiar with prices in your area, get three or four bids, or quotes, from different subs before selecting one. Use a written contract with all subs. As you look for bids, remember my policy to never pay anyone by the hour. If I did, I would have to be on the job site with them the whole time they are working.

You may use the examples on pages 58 and 61 or an attorney can provide you with one. Subcontractors often have their own contracts. At any rate, use one. Don't trust anyone's memory when it comes to dollars. Spell out your specifications thoroughly in your plans to be sure your bids are comparable and that all subs are bidding on exactly the same work.

Asking for Bids

Here is what you should expect when getting bids for various jobs.

Plumbing: Plumbing bids should include all plumbing fixtures right down to the toilet seats. They will not include accessories such as toilet paper holders. If colored fixtures are to be used, specify color and brand. Plumbing showrooms are your best bet for the selection of these fixtures. Magazines and brochures don't tell you enough and often don't give prices. Most plumbing showrooms won't tell you the wholesale price, but you'll be paying list anyway, as the plumber makes a profit on each fixture and it's included in the bid. Don't make an issue of this. The small profit in the fixtures is one of the plumber's sources of income and he earns it.

HVAC: Your heat and air-conditioning contract should include vents (generally fan-powered) for the bathrooms, clothes dryer, stove, and range hood.

Electrical: Electrical bids should include all switches, wiring, receptacles, circuit breakers and their respective panel boxes, a temporary service box and installation, saw service, wiring of all built-in appliances, and installation of ovens and ranges, furnaces, heaters, and air conditioners. Electricians in many areas do the rough wiring for phones and Internet.

Don't forget that the utilities must be connected. Exactly who is responsible for running water lines, sewer lines, and electrical hookups will vary with each subcontractor involved. Get the responsibility pinned down when you are hiring the subs, then follow through to be sure it is done properly.

Scheduling Your Subs

Try to schedule your subs to fit into the sequence of events outlined in chapter 8. This won't always be possible, and one sub can hold up

All subcontractors should be responsible for obtaining inspections from the building department, but make sure they do this or you will have to do it yourself. Lack of inspections can cause delays. Proceeding without inspections can be troublesome and expensive. If, for example, you put up drywall before the wiring is inspected, you could be made to tear down some of the drywall for the electrical inspector. This is not likely, but the inspector could force the issue.

the process. This is why you should check their references and ask whether their previous work has been completed on time. Reliability is as important as quality of work, and in some cases more so.

Working with Subs

If you get along well with everyone at all times, you may not need the next few paragraphs. But if you occasionally run into conflicts, keep reading. The fault — sometimes — may be yours.

At this point you've selected your subs. You've checked them out and are satisfied that they are honest, trustworthy, and experts in their fields.

Subcontractor's Contract

Subcontractor _____

Address _____

Builder _____

Date _____ Plan No. _____

W. Comp. Ins. Co. & Agent _____

Certificate No. _____ Expiration Date _____

Location of Work _____

Total Price for House $_____

Terms of Payment _____

Work to Be Performed and Materials to Be Supplied:

_____

_____

_____

_____

_____

Signed (your name) _____ Date _____

Signed (subcontractor) _____ Date _____

Now let them work. Don't try to supervise the hammering of every nail or the placement of every stud. These guys are professionals and they know more about their trades than you do, and probably, if they came to you well recommended, they take pride in their work. Let them do it.

And, more emphatically, don't try to tell your subs their jobs just because you have read this book and a few others. You'll get good work out of your subs if they understand that you realize they know their jobs and that you're depending on them for good advice and quality work.

When a sub's work is completed, when the work looks good, and when the relevant inspections have checked out, make sure to pay the contracted amount promptly. A hearty thank-you is also in order. Subs who are treated right throughout the job and afterward will do a better job for you, and they'll come back when you build your next house. And you will probably build another!

Paying Your Subs

When you sign your contract with your carpenter, you will agree on a contract price for the work. It is usually based on X number of dollars per square foot of heated area and X number of dollars per square foot of under roof, such as in the garage. Prices will vary with the area, unions, and the complexity of the job.

Work out a schedule of payment with your carpenter. The carpenter and some of the other subs may require draws, or partial payments, as work progresses. This should be agreed upon before work begins. Don't be shy about it. They are accustomed to discussing such matters.

Never pay a sub for work not done, for work that is incomplete, or for an unsatisfactory job. Never pay in advance.

It is all right to pay a draw, but never pay for more than the work already done. If, for example, your carpenter says he is 50 percent finished framing, but he has only framed the floors and walls and hasn't finished the ceiling joists, roof framing, sheathing, or bridging, he isn't 50 percent complete. He is nearer 40 percent complete. Pay him no more than 40 percent of his contract price (bid).

Plumbers and electricians usually get 60 percent of the total contract price when their rough-in work has been completed and inspected. Heating, ventilating, and air-conditioning rough-in payments depend on the installation of equipment such as furnaces. If payment is just for ductwork and some low-voltage wiring, 20 percent of the total should suffice. If a furnace had to be installed during rough-in, add another 10 percent. Work out the arrangement with the sub before he starts. Subs almost always would like to get more money up front than they have in the job. Be sure there is enough money left in the total bid to complete the job if one of your subs goes broke while you are still building. It has happened. You don't want to be stuck paying more to complete his job. You'll be covered better if you don't overpay him on his rough-in.

Brick masons and painters are about the only other subs who will require a draw in progress. You will have to use your best judgment as to how much of the job is done. Again, don't get ahead of them in paying.

Make sure inspections by your county or city are completed and the work is approved before you make any payments at any phase of construction, other than partial draws. This is your assurance that the job has been done, and done properly.

I seem to be saying the only way you'll get your sub to complete the job is if you owe him money. In some cases that is true, but in others it is only partially true. Some subs would finish regardless. Often the issue is that subs have more than one job going at one time, so your main objective is to get your job finished before one that was started after yours.

Chapter 7

Suppliers

THE SUPPLIERS THAT YOU WILL BE BUYING FROM are listed in this chapter by the type of products they sell. Some sell several different products, so you can shorten your shopping time when you patronize them. All of them will require a credit check on you. Your suppliers will also want the name of your lender. This lender reference is the key to your getting credit with them. Obviously, if your credit is strong enough for the lender, it's strong enough for the suppliers, so often just the lender reference is enough.

Finding Your Suppliers

The suppliers you most likely will be using are:

- ◇ Sand and gravel company (for driveway and backfill)

- ◇ Brick company (for face or decorative brick)

- ◇ Concrete block and brick company (for mortar mix and the building blocks for foundations)

- ◇ Concrete supply company (for concrete for basements, garages, footings, and driveways)

- ◇ Lumber company or home center (for framing lumber, nails, windows, doors, roofing, siding, paneling, and interior and exterior trim)

- ◇ Flooring company (for vinyl, wood, and tile flooring and for carpet)

◆ Cabinet shop (to fabricate custom cabinets. Most lumber companies also sell both ready-made and custom cabinets.)

◆ Countertop company (for granite or marble countertop, hearth, vanities, and so on.)

◆ Lighting fixtures supplier (for light fixtures, bathroom fans, exhaust fans. The firm will do a complete count of your lighting needs and help you stay within a dollar figure set during estimation.)

◆ Paint store (for wallpaper. Paint is usually purchased by your painter and included in the price per square foot.)

◆ Appliance store (Some lumber companies and lighting centers carry appliances.)

◆ Insulation company

◆ Tile company (for ceramic tile, marble, and any decorative stonework)

◆ Drywall company (Sheetrock is the most common brand. Some lumber companies and home centers also sell drywall.)

◆ Any other specialty type suppliers that you may need for certain items not carried by one of the above.

Large home centers or lumber companies may carry most of your needs. Some carry everything from bricks to wallpaper. So long as their prices are competitive, that's fine. In some cases, however, the larger stores will probably have quantities of basic material in stock, such as tiles or carpeting, but may not have as wide a selection of choices as a specialty store. It's worth it to look around to see what your options are.

This is my recommendation: When you find a good lumber company or home center, give your plans to the "contractors" sales department and let them give you a list of all the items the company can provide and a complete cost list. I have never found a company that wouldn't do this. Some will do a complete takeoff (material list) of all the lumber and other materials you will need.

You will get a material list from your lumber company of the number of studs, floor joists, windows, and rafters. It will be fairly accurate, so don't be afraid to use it. Remember, the company is eager to sell to you, and this is part of its service. Company personnel can also explain different products to you and show you new products and ideas. Let all your suppliers aid you by providing take-offs and different product ideas.

Delivery of Materials

Most companies will also assist you in, or be responsible for, delivery, by making certain the materials are delivered on time, but not so far in advance that they may be stolen off your lot. Make sure there's a flat, clear area close to the foundation where the load of framing lumber can be off-loaded. This delivery system has worked particularly well for me when I have used one supplier for all framing and as much of the rest of the supplies as I could. In such a case I have permitted my carpenter to order needed materials from that supplier alone. The supplier has aided me in keeping an eye on the additional materials ordered.

Buying at Builders' Cost

"Ask and ye shall receive." If you don't, you won't. Be sure to ask whether the quoted price is the "builders' price." Tell the supplier that you are a builder, because you are. If you want to add "company" to your name (that is, John Smith Company), fine. If it makes you feel better, do it. It costs nothing and changes nothing legally (the credit will still be in your name). If you want to incorporate (form a corporation), talk to an attorney. I don't think you'll find it necessary.

Paying Your Suppliers

Some suppliers offer at least 30-day terms so that you pay the stated amount within that time without penalty. Some independent stores may also offer a small discount if your bill is paid by the tenth of the month following purchase. For example, if you buy supplies on June

20, you must pay by July 10 to receive the discount. If you buy on June 1, you still have until July 10 to pay and get the discount. I try not to order too much toward the end of the month and am willing to wait as much as a week, if I can, to get an extra 30 days to pay. Not all suppliers are as generous as this, and their terms will vary. Check with your suppliers on what terms will be offered.

By getting favorable terms and extra time in which to pay, you can be sure that your construction draw will cover the bill. You will use construction draws to pay for all your supplies. You should seldom have to use any of your own money. If you do, your construction loan should reimburse you.

Construction draws are based on labor completed and materials in place, not stored or sitting on the job site, although some lenders will pay for materials on the site. For example, if you order brick for the whole house (a brick veneer home) at the start of the job, you'll have to pay for the brick that's going to be used in the veneering before the lender advances the dollars for veneering. Instead, you should order the brick in two stages, first for the foundation, then for veneer when you're ready.

Waivers of Lien

Whenever you pay a subcontractor or a supplier, you should always have them sign a waiver of lien. A waiver of lien means that the person or company has been paid for a particular service or supplies and has given up their legal rights to seek more money for that service or supplies. Laws differ from state to state, so check with a lawyer or title insurance company in your area for information on the forms and requirements concerning waivers of lien.

Bookkeeping

Bookkeeping is very simple when you are building only one house. Open a separate checking account to handle the construction of the house and pay all bills, no matter how small, by check. Save all receipts, no matter how small, and waivers of liens. Your lender, tax advisor, and you will need them. Code each check you write to one of the appropriate categories on the estimate spreadsheet and list the amounts on a separate sheet of paper or separate spreadsheet until you have a total of all of the money spent on that category. Then enter the total amount in the adjusted cost column of the estimate spreadsheet and compare the adjusted cost to the original estimated cost.

You can make as many spreadsheets for sub-categories as you need. For example: Foundation costs may be made up of several costs such as sand, fill dirt, steel reinforcing, forms, and so on. Simply start a new spreadsheet, label it "Foundation," and change the list of costs by typing in what you need. The pre-typed categories can be overwritten.

Chapter 8

Building the House

LOOK AT HOW MUCH YOU HAVE ACCOMPLISHED. You have your building lot, you've arranged your loan, and you have a house plan you can live with and are able to finance. You've completed a vast amount of paperwork that includes any necessary permits and insurance policies.

You have located most of your subs and have contracts with them. You've visited various supply houses and worked out your accounts for bricks, concrete, and lumber.

Congratulations. You've reached the day you thought might never come. You're ready to start building.

What is the proper sequence of steps in building the house, and how long will each take? Let's make a list of them.

1. Staking the lot and house: 1–3 hours
2. Clearing and excavation: 1–3 days
3. Ordering utilities, temporary electric service, and a portable toilet: 1 hour
4. Footings (first inspection must be made before pouring): 1 day
5. Foundation and soil treatment, then foundation survey: 1 week
6. Rough-ins for plumbing, if on a slab, and inspection: 2–4 days
7. Slabs, basement, and garage: 1–2 days
8. Framing and drying-in: 1–3 weeks
9. Exterior siding, trim, veneers: 1–3 weeks
10. Chimneys and roofing: 2 days–1 week

11. Rough-ins (can be done during steps 9 and 10): 1–2 weeks
12. Insulation: 3 days
13. Hardwood flooring and underlayment: 3 days–1 week
14. Drywall: 2 weeks
15. Priming walls and pointing up: 2 days
16. Interior trim and cabinets: 1–2 weeks
17. Painting: 2–3 weeks
18. Other trims, such as Formica, ceramic tile, vinyl floors: 1 day–1 week
19. Trimming out and finishing plumbing, mechanical, and electrical and hooking up utilities: 1–2 weeks
20. Cleanup: 2–3 days
21. Carpet and/or hardwood floor finish: 3 days–1 week
22. Driveway (if concrete, can be poured anytime after step 14): 1–3 days
23. Landscaping: 1–3 days
24. Final inspections, surveys, and closing of construction loan and interim loan: 1–3 days
25. Enjoying your home: a lifetime
    Note: Steps 2 and 4 can be done by one sub. Steps 3 and 4 can be reversed.

The Steps Explained

Let's take a closer look at those steps and clear up any details you should know about them.

STEP 1: Staking the Lot and House (1–3 hours)

The number one problem that occurs when building is the placement of the house on the lot. Staking is vitally important in the early stages of the building process. Since this is most likely the first house you have ever built, I recommend hiring a registered surveyor/engineer to stake both the lot and the position of the house. If you can, use the surveyor or survey company that did any previous survey of the property. It is less expensive for a surveyor to resurvey a property than to do an initial survey because it takes less time.

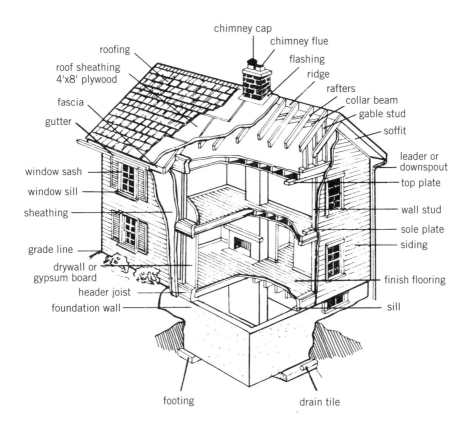

chimney cap
chimney flue
roofing
flashing
roof sheathing
4'x8' plywood
ridge
rafters
fascia
collar beam
gutter
gable stud
soffit
window sash
leader or downspout
window sill
top plate
sheathing
wall stud
sole plate
grade line
siding
drywall or gypsum board
header joist
finish flooring
foundation wall
sill
footing
drain tile

Surveying is an important function, as houses have been placed in violation of certain setbacks or restrictions and builders have placed houses straddling property lines and have had to tear the foundations out and start over. Best to be safe. If a surveyor does this, he is responsible for the expense of making corrections.

The stakes for the lot itself may have been moved or torn down since you purchased it. A surveyor will check this. The cost to re-stake a lot and stake a house should be minimal, depending on the complexity of the house and the terrain. You, of course, will want to be present for the staking or placing of the house to be certain it faces the direction that you want. If possible, meet with the surveyor at the site before he begins working. Or you could put some stakes in the ground over a weekend to indicate the approximate location and direction you desire. Then let the surveyor do it accurately and you can inspect it when it is done.

When the lot is cleared and the basement — if you have one — is dug, you may want the surveyor to re-stake the house. On his first visit he usually will put in offset stakes. Offset stakes mark the original four or two corners but are offset as much as 25 feet to the side so they won't be disturbed during clearing and excavation. Then it's a quick job for him to find the exact locations on his second visit.

Usually your house should face in the same direction as other houses on your side of the street. It doesn't have to, especially if your house is on a large lot, unless the building code demands it.

If your lot slopes more than 3 or 4 feet, you may need a topographical plat (called a "topo") from your surveyor. Shop for the best price. Topos are expensive, but you'll need one so that you can fit the house to the slope and be certain that water will go around the house as it drains off the grade.

Water is, has been, and always will be the biggest problem that interferes with the construction of a building. As nature's strongest force, it goes wherever it can, taking the shortest route. It is stronger than anything man can make. Keep in mind that gutters help get rid of water; they don't control it.

The house should be positioned first on a map of your lot, then that position should be staked on the site by you and your carpenter, surveyor, or footing contractor. The surveyor is best for the job. An architect or designer can position the house on the lot on paper, considering water as well as solar energy, wind, light, and the location of other houses. Make sure you convey your preferences to the person positioning the house before they start working on your job. Here are few things to consider:

◇ Interior light: A north-south facing house will be darker than an east-west facing house.

◇ Water flow: How will it affect landscaping and basement drainage?

◇ Other houses on the street: Setback requirements are determined by local zoning and deed restrictions, but they are also influenced by your neighbors' houses. How are other houses oriented on their lots?

◇ The street itself: Should the house be parallel with the street? What if the street curves? What about a corner lot?

◇ Privacy on the front, back, and sides: Think about what your windows will overlook and what your neighbors might see from their homes. (You may want to change window placement based on this factor.)

◇ Solar orientation: To get the most benefit from the sun's energy, homes should have as much surface as possible facing south, with broad expanses of glass on that side and a minimum of doors and windows on the north side. Homes may also be designed to allow for roof solar collectors or other solar energy devices. (Find out more from companies selling the technology.)

◇ Minimum setback and side-yard requirements: Make sure you are within the legal boundaries of your property.

I don't want to frighten you with so many considerations. I have built many houses for my family, and I still have to think through these decisions. Positioning the house is a very personal decision, one you should make when you first buy your lot. If you are stumped and can't decide for yourself, and you don't have an architect or designer involved with your plans, spend the money and get the advice of an architect for that particular decision.

STEP 2: Clearing and Excavation (1–3 days)
Clearing the lot includes removing trees, brush, rocks, roots, and debris from where the house will sit, and usually 10 feet or more around the foundation, thus allowing space for tractors, fork lifts, and trucks working at the site. Obviously, the more area to be cleared, the more it will cost. Removing big trees is time-consuming and expensive. Rocks may have to be blasted. If you want trees to be

cut into firewood, the crew will charge dearly for doing the job. Best to have the good, manageable-sized logs cut down to 10 to 12 feet in length and piled at the side of the lot for you to cut up at your leisure.

Your best source for finding a sub for clearing and excavating is by word of mouth or in the Yellow Pages. Get a contract price for this work. It may cost a little more, but you will be assured of not having your first cost overrun. If a basement is to be dug, your sub must have and know how to use a transit. You may also want your surveyor/engineer to oversee this digging to make certain it is the proper depth. Some or all of the dirt removed from the basement site might be put aside for later backfilling and landscaping. Topsoil should be separated out to be spread later for the lawn or garden.

Your contract price should include hauling all trash, such as stumps, branches, and rocks, to a suitable landfill. I don't advise burying trash on your lot, as these stump holes tend to form unsightly depressions as the material settles or rots. Some areas ban stump holes. Of course, if the distance to a suitable landfill is prohibitive and you have enough land, a stump hole may be your best choice.

You may want to put one or more loads of stone on your driveway so that supply trucks can drive in during wet weather. The best stone for this is unwashed crushed stone. It has a powdery substance created in crushing the stone that will harden after getting wet. You also may need to put drainpipes in at the roadside if they are required by either ordinance or common sense. They allow roadside water to flow under the driveway and prevent the stone from washing away.

STEP 3: Ordering Utilities (1 hour)

When you purchased your lot, you determined what utilities were available and how much they would cost. Now it's time to make plans for a couple of months down the road with a few phone calls or a visit to each utility. Pay all fees and complete any necessary forms. Arrange for temporary electric service for your subs. Usually your electrician is responsible for installing the temporary electrical panel box and having it inspected, but you will have to apply for the service from the utility. This usually can be done over the phone.

Wells and septic systems, if used, can be installed now, and it is best to have this work done at this time. County or city health inspectors may be required by code to determine the location of these. Tell them your plans for such things as gardens or driveways, or which trees you hope to save, to guide them in their decisions.

If no temporary source of water is available, such as from a house next door, you will have to have the well dug and temporarily wired for your brick masons, who will be needed shortly, or they will have to truck in their own water.

I also recommend, and some locales require, a portable toilet on the job site. Sources for renting these can be found in the Yellow Pages under Toilets-Portable.

STEP 4: Footings (1 day)

The footing is the base of a structure. It is a mass of concrete supporting the foundation of the house. It can be poured into wooden forms or in trenches. It must be below the frost line or it will heave when the ground thaws and freezes. In the northern states and higher elevations of any area, this line may be 4 or more feet below grade level. This is one reason there are more basements in northern climates. If you have to be several feet below grade level for your footing, and thus have several feet of foundation already in place to get back up to grade level, only excavation and a concrete slab are needed for a basement. Local codes will clearly state the requirements for footings in your area, and your subcontractor should know the code.

I have a footing sub who stakes, clears, excavates, digs, and pours footings. For your first house, I recommend that you find one who does the same. The cost will be comparable. The footing is probably the most important part of the house. If it settles or moves, so will your house. If it is not done according to the dimensions of your plans, you will have to change the plans to accommodate the footing or do the footing over. I recommend the former if the situation arises, unless the deviation is too severe.

After your foundation walls are up, put in a footing drain. Your code may require this. The drain can be connected to a dry well, storm sewer, or any other approved means of getting rid of the water. In some places, it can simply drain into your yard.

As a rule, footings are better today than they were 100 years ago. Well-built houses of today will probably last years longer than those built long ago. Technology has improved materials such as concrete, and our knowledge of how to use them has increased. I say this to help ease your mind about this important step.

Building inspectors usually check the locations of footings before they are poured to make certain they are deep enough and resting on undisturbed earth. Don't complain about this inspection — it could save you thousands of dollars if it means that you will avoid some future problem.

STEP 5: Foundation (1 week)

Foundations can be made of brick, concrete block, or poured concrete. Stone foundations generally aren't built anymore, as they aren't as strong as the others. Stone can be applied as a veneer just like brick, for aesthetic purposes, and you would be wise to use it only as such. Local codes may actually prohibit stone as a foundation load-bearing material.

Your masonry contractor needs to be one of your more experienced and reliable subs. Next to your carpenter, he is the most important. Your carpenter probably can recommend a good mason, as he usually starts his work soon after, if not directly after, the mason is finished. He knows the good ones. I'm sure he's had to follow bad ones and probably remembers having to shim walls or compensate in some other way for an out-of-square foundation. Houses can and should have square corners.

The foundation wall for any type of house needs to be high enough so that water will be diverted away from the house by the final grade of the soil around the house. It must also be high enough so that the wood finish and framing of the house will be at least 8 inches above the finish grade and thus protected from soil moisture.

A crawl space should be at least 18 inches high so that you can crawl beneath the house annually to inspect for such things as termite damage. The crawl space walls should have screened openings for ventilation. If you want a full basement, your foundation walls should be high enough so that you have at least 7 feet 4 inches of headspace in the finished basement.

If you are in doubt about the foundation height, consult an engineer. Usually you, with the help of the carpenter, mason, excavator, or anyone who can use a transit or a level, can determine the needed height. If the lot is almost flat, the job is simple. It becomes tricky when the lot is steep or has opposing grades. Experienced contractors make certain the foundation is high enough at the highest point of the outline of the foundation wall, and use that highest point as the control point.

The finished foundation should be waterproofed from the footing to the finish grade line. I recommend hiring a professional waterproofing company for this. Companies are listed in the Yellow Pages under Waterproofing. Don't let one of your regular laborers do it in his spare time, if he offers to. A professional company will stand behind its work.

Also, depending on your locale, you may need to have the soil treated for insects and pests, particularly termites. Hire a professional. This job is done after the foundation is in, but before any concrete is poured for either the basement or the garage.

It may come as a surprise to many first-time home builders that the foundation is poured or formed with concrete blocks, then holes are punched in it for such things as the water supply and the sewage outlet, the pipe is placed through the hole, and the space between the pipe and wall is patched. This is the easiest method for getting a tight, waterproof fit.

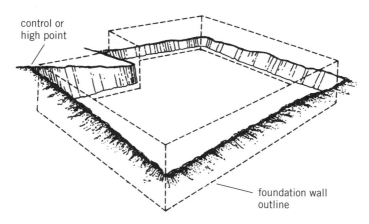

How to establish the depth of excavation

STEP 6: Rough-ins for Plumbing (2–4 days)

If you have a basement with plumbing or if you are building the house on a concrete slab (as opposed to wooden floor joists), once the foundation is in and backfilled and tamped (packed down), and the soil treated, your plumber needs to install the sewer line and the water pipes that will be under the concrete. Also, any wiring that will go under the concrete needs to be placed in a conduit and roughed-in. Most wiring, though, can be run through the stud walls and ceiling joists to any given point.

Your treatment company may want to wait until the rough-ins are completed before treating the soil so it won't be disturbed by digging in the plumbing lines. Ask about your company's policy.

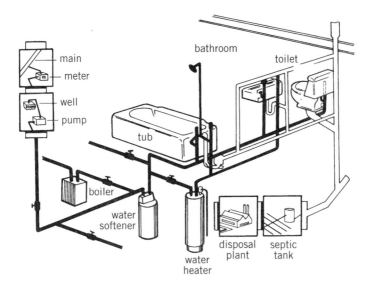

The plumbing system of a home. Note that the supply system pipes are black, the drainage system's are white.

STEP 7: Slabs — for Heated Areas (1–2 days)

Many locales require slab perimeter insulation. The most common perimeter insulation is extruded polystyrene foam board (poly), which is available in various thicknesses. The board extends from the top of the foundation to at least 12 inches below finished grade. I recommend a 1-inch or thicker board, even if it is not required by codes.

I also recommend using a plastic barrier of 4–6 mil thickness under the concrete to prevent moisture seepage. A 6-by-6-inch #10 wire mesh should be placed in the concrete to reinforce it.

The top of the slab should be at least 8 inches above the finish grade. Your sub should put down a base for the slab, tamping down gravel or crushed stone to form a layer 4 to 6 inches in depth. The poly goes down on this just prior to pouring the concrete. If you cannot cover the entire area with one sheet of poly, any joints of the poly should overlap by 4 inches and be sealed with silicone caulk. The wire mesh is laid on top of the poly. Call for an inspection before pouring concrete if your code requires it.

A good concrete sub will do all of this. I stopped checking my slab pourings when I found a terrific concrete subcontractor.

For garage slabs, with proper backfilling and tamping, you don't need the wire mesh, but if you believe the added strength is required, use it. I recommend an expansion joint of fiberboard around the perimeter. Garages are subject to extreme temperature changes, and concrete expands and contracts with those changes. The expansion joint permits this expansion without cracking the concrete. Both this slab and the house slab should be at least 4 inches thick. Be sure your sub thickens all slabs wherever they will be carrying load-bearing posts or walls. Codes differ on the additional thickness required, but it is often the thickness of the footings.

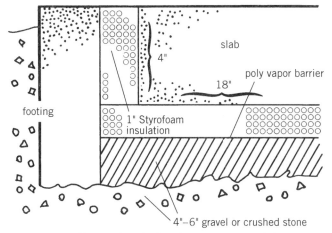

Proper insulation for a concrete slab

Framing Package

15	pieces	2 x 6 x 12 treated pine
10	pieces	2 x 4 x 12 treated pine
20	lineal feet	2 x 2
250	lineal feet	1 x 4
18	pieces	2 x 8 x 12
80	pieces	2 x 8 x 14
80	pieces	2 x 8 x 16
55	pieces	2 x 10 x 12
100	pieces	2 x 10 x 14
18	pieces	2 x 10 x 16
18	pieces	2 x 10 x 18
2	pieces	2 x 10 x 20
14	pieces	2 x 6 x 14
45	pieces	2 x 6 x 16
80	pieces	2 x 6 x 20
800	pieces	Studs, 2 x 4 x 93
150	pieces	2 x 4 x 14
50	pieces	2 x 4 x 10
175	pieces	1/2-inch CDX plywood
24	quarts	Construction adhesive
10	rolls	#15 felt
60	pieces	Asphalt-impregnated sheathing, 1/2 x 4 x 8
300	pounds	16d nails, coated
150	pounds	8d nails, coated
10	pounds	8d steel cut masonry nails
25	pounds	7/8-inch galvanized roofing nails
50	pounds	1 1/2-inch galvanized roofing nails
10	pounds	16d galvanized finish nails

STEP 8: Framing and Drying-In (1–3 weeks)

If you have a good carpenter, this is the end of this section of framing. And if that sounds too simple, wait and see. You need only order the lumber, the windows, and the exterior doors, and in a few weeks you'll have a house (or at least something that looks like a house). I have included here a typical order list for framing and drying-in, which means making the house secure from rain. Rain or snow during framing is not desirable, but it seldom does much damage beyond an occasional warped piece of lumber. But once the house is dried-in and the windows and doors are in place, the work inside can progress regardless of the weather.

Various types of windows

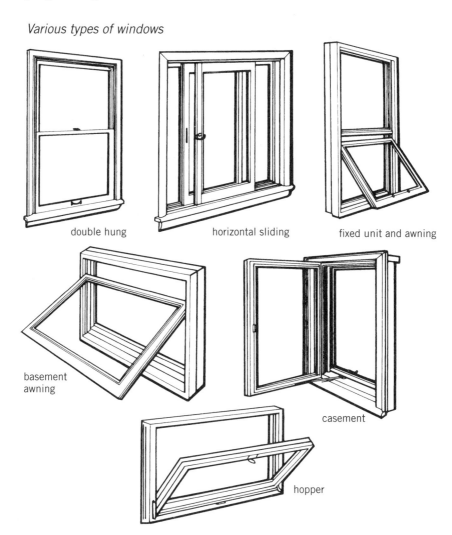

double hung

horizontal sliding

fixed unit and awning

basement awning

casement

hopper

You will want to check with your carpenter about any problems with materials. Because it is impossible to estimate exact needs in materials, some ordering will have to be done during the course of construction.

You could give your carpenter permission to order what he needs and then tell the supply house what you have done. The decision is up to you. I do it.

The drawings of typical framing (shown below) are to make you familiar with some of the construction terms.

When the framing is completed, order cabinets, bookcases, and vanity cabinets, if there are any. The sales representative can measure space for them on the job.

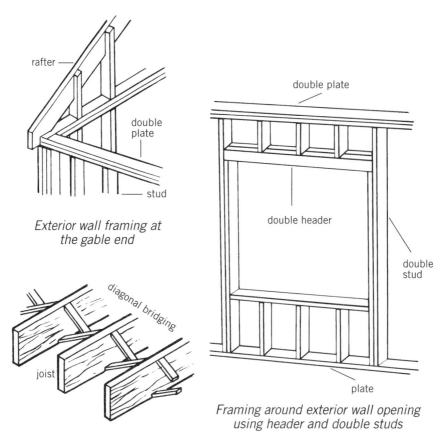

Exterior wall framing at the gable end

Diagonal bridging of floor joists

Framing around exterior wall opening using header and double studs

STEP 9: Exterior Siding, Trim, Veneers (1–3 weeks)

This phase of construction is carried on while work progresses inside and should be done before roof shingles are installed. Masonry chimneys are installed after siding or veneer. Veneers such as brick should be installed before final exterior trim (boxing) is added. At completion of this step, you are ready for exterior painting.

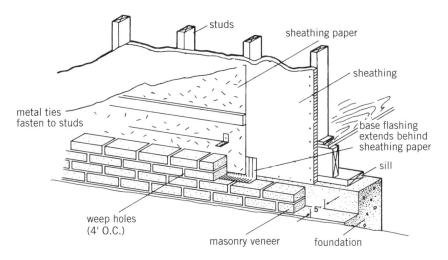

A wood frame wall with masonry veneer. Note shelf on foreground of foundation, to support bricks.

STEP 10: Chimneys and Roofing (2 days–1 week)

Chimneys should be built before the roof is shingled. This will allow placement of sheet metal flashing around the chimney for waterproofing and will also avoid damage to the shingles. A prefab fireplace and flue would also be installed at this time. Roofing follows completion of the chimneys.

STEP 11: Rough-ins (1–2 weeks)

All electrical, plumbing, heating and air-conditioning, phone pre-wires, cable or satellite TV lines, stereo and intercom, Internet, and burglar alarm systems should be roughed-in at this time or anytime after step 8 is completed. This does not mean that these units are installed at this time — only the wiring or plumbing for them. Inspections are needed when this step is complete.

STEP 12: Insulation (3 days)

Consult with your local utility company on the insulation you need in order to qualify for their lowest rates. Some locales require an inspection of insulation by both the utility and the building inspection department when it is completed, and before it is covered with drywall or paneling (or plaster, if anyone still wants plaster walls).

STEP 13: Hardwood Flooring and Underlayment (3 days–1 week)

I find it easier and neater to install unfinished hardwood flooring and vinyl or carpet underlayment before I have the drywall installed. It can also be done afterward. Never install finished flooring before the drywall goes up.

STEP 14: Drywall (2 weeks)

Most residential interior walls are finished with ½" x 4' x 8' gypsum wallboard sheets called drywall (Sheetrock is a brand name). For baths and other moist areas use a waterproof board or paint the drywall with an enamel paint, even before wallpapering.

Figure on 3½ or 4 times the square footage of floor area for the total square footage of wallboard to be used. Your drywall sub can give you a price based on a square foot charge. Some will give a bid that includes materials. On your first house you may want to go this route.

STEP 15: Priming Walls and Pointing Up (2 days)

After the drywall is finished and before the interior trim is applied, I prime all walls and ceilings with a flat white latex primer. My two painters do this with a spray gun and in one day can complete a house that measures 3,000 square feet. Priming reduces the finish painting time considerably, which saves you money. With no trim installed, only windows need to be protected during spraying. If all of your woodwork (trim) is going to be painted instead of stained, priming the walls by spraying can be postponed until the painting stage.

When the walls and ceilings are primed, the slight imperfections that sometimes occur in finishing drywall will show up. At this time you can have your drywall subcontractor touch up these places so that the walls and ceilings will be ready for final painting (this is called pointing up). Pointing up can also be done just after the interior trim is finished in case the walls get nicked during this process. I have done it both times, and at no extra charge.

STEP 16: Interior Trim and Cabinets (1–2 weeks)
Doors, moldings, cabinets, countertops, and shelves are installed at this time. This includes kitchen cabinets, bathroom sink bases and medicine cabinets, and all other built-in cabinets such as bookcases. Cabinets that were ordered after the completion of step 8 should be ready for delivery and installation. The cost of installation should be worked out in advance between you and either your carpenter or supplier (or cabinetmaker if you use shop-built cabinets). Interior trim labor usually includes standard trim and moldings. Any special molding or trim work, or any paneling, should be discussed with your carpenter in the planning stages to determine the extra cost of installation. Your millwork supplier, usually the lumber company, can do a complete material takeoff for your trim and discuss the costs of extra items.

STEP 17: Painting (2–3 weeks)
You are ready for final painting inside. Your exterior painting can be delayed until this time, too, to save your painter from hopping back and forth from job to job. Work this out when you first discuss painting with your sub. You don't want to leave the exterior trim unpainted or unstained too long, as it may warp or get moldy.

STEP 18: Other Trims (1 day–1 week)
It's time to install vinyl floors and ceramic tiles. Wallpaper can be hung at this time or delayed until after you move in. The flooring can't be delayed, because the plumbing cannot be completed until the floors are finished.

STEP 19: Trimming Out (1–2 weeks)
It's time for the plumber to finish his work. This is called trimming out or setting fixtures. Some of the fixtures he installs must be wired, so he needs to finish before the electrician can finish. Also,

your heating and air-conditioning must be completed before your electrician finishes his work. Next comes your electrician to install switches, receptacles (sometimes called devices), light fixtures, and electrical appliances, such as the oven and range. He will also wire the electrical apparatus that has been installed by your plumber and heating and air-conditioning subs.

STEP 20: Cleanup (2–3 days)

Both inside and outside cleaning can be started. The bulk of outside trash, and an incredible amount of it that comes from inside, can be picked up by truck and hauled away. (In case you're wondering, dumpsters are the responsibility of the general contractor. They don't hold much, however, and are best suited for renovation work where trash containment is more critical. On new construction sites where there is no landscaping in place, it's cheaper to bulldoze all the trash and truck it away at the end of the project.) You can do the inside work yourself or hire a professional crew. Be sure to add the cost to your estimate.

STEP 21: Carpet and/or Hardwood Floor Finish (3 days–1 week)

Hardwood floors should be finished before any carpet is installed, because of the sanding required before the floors are stained and sealed with polyurethane. The dust from sanding usually is controlled, but some could permeate the rooms with carpeted floors. Because of the dust, you may want to finish the floors before step 20, cleaning up. I've done it both ways and found advantages to both.

STEP 22: Driveway (1–3 days)

You should keep heavy trucks off newly laid concrete or asphalt drives for a period of time. Try to time it so that you have received all heavy shipments of materials, hauled off all heavy loads of trash, and finished with any heavy equipment coming to do work. Realize that if you wait until the end with the driveway, you'll at least have a moving van on it. Concrete can support a moving van a week after the concrete is poured, but asphalt can't. If you use asphalt, wait until after you move in to pave. Put down a good stone base on the drive before moving in. My asphalt-paving contractor insists on this and puts down the stone base and then redoes the stone (or dresses it up) before paving it with asphalt several days after the moving van leaves.

STEP 23: Landscaping (1–3 days)

This job can be put off until after you move in, depending on the requirements of your lender. Such a situation may arise due to weather or scheduling problems or lack of money due to other cost overruns at the end of construction. You might get by with just grading and seeding or mulching disturbed wooded areas.

STEP 24: Final Inspections, Surveys, Loan Closings (1–3 days)

After the completion of the actual house (not the driveway and landscaping), all final inspections from the county or city for building, electrical, mechanical, and plumbing should be made. Also, the lender will make a final inspection of the house at this time and may require driveways and landscaping to be complete before it will disburse the balance of the construction loan.

When approval is given, your loan officer will coordinate the necessary paperwork and schedule the refinance or modification of the construction loan. The lender will most likely require a final survey to be sure no additional structures or additions have been placed on the lot in violation of deed restrictions or zoning. Generally, your lender will order this final survey and any other necessary documents required for closing.

The actual loan closing may take only about 15 minutes, depending on the efficiency of the lender. One document you must remember to obtain (your lender will remind you of it) is your insurance policy. It will need to be converted into a homeowner's policy prior to closing. This merely takes a phone call to the agent who issued you your builder's risk (or fire) policy.

STEP 25: Enjoy your new home!

Chapter 9

Add On, Remodel, or Tear Down and Start From Scratch

THERE ARE THREE THINGS YOU CAN DO to an existing house to increase its value and increase its livability:

1. Add on
2. Remodel
3. Tear it down and build a new one (the Extreme Makeover approach)

With all three approaches, remember that "general contracting is general contracting," and you will always save money being your own house contractor, no matter what the size of the project.

Obviously, you can, and people often do, combine an add-on project with a major renovation, but I will treat them separately, for there is a difference in the scope of each job. In both cases, you will save even more (as a percentage of cost) by acting as your own contractor than on new construction. With the third option, other than the demolition, the process is the same as building on an empty lot.

Building an Addition

If you would like more room and don't want to move, you can add to your present home. It could be easier than you think, as you may already have a head start toward additional space.

There are two basic ways to add to a home. One is to make habitable an existing unfinished area such as a basement, garage, screened porch, breezeway, or attic. This is the least expensive way because you already have the basic room with a foundation and some or all of the outside walls. The other and more expensive way is to build on to your house. If you are considering this, check to see that you have room on your lot without violating required setbacks and that you meet any subdivision or deed restrictions.

Whether you finish an existing space or add to your present structure, the procedure will be the same. Start at the beginning of this book and treat the addition as a small home. The only things that will be different from building a house are that you already have the land and the costs will be somewhat different. Don't skip any of the applicable steps I outlined earlier for building a house.

Don't add too much value to your home. Sooner or later, all houses need to be sold, and you don't want to have the most expensive house in the neighborhood when you put it on the market. It's a real estate fact of life: The most expensive house in the neighborhood is very difficult to sell.

How to Estimate Costs

The basic difference in estimating an addition against a complete home will be the increase of some square footage building costs due to the smaller size of the job. Subs may want a few more dollars for their labor because they could be earning more on a larger job for almost the same amount of time. Renovation costs are higher because it takes the same time to bid on the job and set up the worksite, but the work usually goes more slowly as there are different issues to worry about, such as working around existing walls, finding unforeseen problems during the teardown phase, and so on.

The general contractor that you won't be hiring because you're doing it yourself would also have charged a greater percentage for profit and overhead. As I mentioned earlier, the National Association of Home Builders recommends that professional builders aim for a 50 percent gross profit margin. The savings you realize by being your own general contractor will more than offset the increase in labor costs from subs.

You should make a list of all the materials the project will need (using the information in chapter 3), then obtain competitive bids as though you were building a house and proceed from there.

How to Finance an Addition

Financing an addition could be easier than financing a new home. You may be able to simply get a home equity loan based on the equity you have in your home. (Your equity is the difference between what the house is worth and what you owe on it.) Quite often you don't even need to give a reason to borrow against that equity because the lender is well protected if you should default. The house already exists and serves as collateral for the loan.

The amount you can borrow varies from lender to lender, but if you have been in your house for at least three years, it is safe to assume you have built up some equity. Typically, you can borrow up to 100 percent of your equity. If you don't have sufficient equity for an equity loan, a construction loan based on the "completed value" of the house may well be the answer.

Financing small jobs can also be done with cash from savings, low-interest credit cards, or even a construction loan. Construction loans however, because they are more costly, are usually for larger jobs than a kitchen or bath redo. Generally they are for major projects that increase the value of the house tremendously. Just as in planning new home construction, your local lender can sit down with you and discuss your lending needs.

Remodeling, Renovation, or Restoration

Over the years, finding inexpensive historic houses to buy for the express purpose of restoring them has gotten harder, except in areas that no one wants to live. It is difficult to increase the value of a home through renovation if it is in an undesirable area. The result is you may have to pay too much for a structure to make renovation feasible. If the price of the house plus the cost of renovation exceed the fair market value (appraised value), it makes little sense to proceed. You must keep in mind that sooner or later all houses have to sell.

There do remain a few pockets of desirable older homes, usually in established neighborhoods close to viable downtown areas, where paying more to purchase is rewarded in the end. But the big bubble of the 1980s and '90s is long gone. If you already own a house in need of renovation, however, or have found an appropriate one to buy, here is how to analyze the situation.

1. Make up a budget.
2. Have the house inspected by a reputable home inspector.
3. Determine what upgrades must be done and what improvements you want to make. Consult the inspector and use the renovation checklist on pages 92–94.
4. Determine costs by using the cost estimate spreadsheet just as you would for a new house, modifying it to reflect the needs of the project.
   Note: You will find that labor costs in renovation are much higher (as much as double) than with new construction or even additions. This is because subcontractors know that there are often hidden costs in renovation such as finding problems behind walls or making things fit into an existing space. They also know that extra care must be taken when working in an existing house. Don't worry, though — material costs are the same for either new construction or renovation.
5. Get an appraisal as to what the house will be worth when the work is done.
6. Make your decision based on the appraisal. Will the added cost make the house too expensive for the neighborhood? This will be indicated on the appraisal. If it does, it will be hard to sell and/or you will lose money. The appraisal may actually indicate a value far below your estimated costs, which could be a function of the neighborhood's housing values dragging your project down or an indication that you didn't do a careful job of estimating.

7. If you decide to go ahead, follow the same steps for general contracting as you would in new construction as far as obtaining financing, scheduling subcontractors, and so on.

Even with small remodeling jobs, it makes sense to act as your own general contractor. After all, why pay a remodeling general contractor to make the same phone calls you can make to the plumber, electrician, tile contractor, cabinet shop, painters, flooring contractor, home center, and so on?

Say you're planning a new kitchen. Make a list of what you want done. Find corresponding subcontractors for each type of job and get bids from them as well as a list of materials they may need you to supply. Enter all those costs on a spreadsheet and see what it costs with you being the general contractor.

Next get a bid for the whole job from a remodeling general contractor, where all you have to do is write the check. See which way is cheaper. Will your homework and scheduling efforts be worth it? I think you'll find out the answer is *yes!*

LIST OF ALL POSSIBLE COSTS FOR TOTAL RENOVATION

	ORIGINAL COST	ADJUSTED COST
Inspection fees, testing (lead, asbestos, radon, and water)		
Design assistance		
Loan closing costs		
Construction or short-term loan interest		
Fire insurance		
Temporary utilities		
Plans and specifications		
Permits and fees		
Clearing, grading, and excavation for an addition		
Footings		
Foundations		
Waterproofing		

	ORIGINAL COST	ADJUSTED COST
Tearing out		
▸ exterior walls		
▸ interior walls		
▸ roof-shingles		
▸ roof rafters		
▸ plaster, drywall		
▸ furnace, radiators		
▸ plumbing		
▸ electrical wiring, etc.		
▸ total gutting		
▸ disposal dumpster fee		
▸ asbestos removal		
▸ lead paint removal		
New framing — lumber, materials, and labor		
▸ walls		
▸ floors		
▸ ceilings		
▸ roof		
▸ additions		
Steel		
Windows		
Exterior doors		
Roofing — labor and materials		
Concrete flatwork		
Exterior trim — labor and materials		
Exterior siding or veneers		
Plumbing — upfitting		
▸ toilets		
▸ bath sinks		
▸ kitchen sinks		
▸ tubs, refinished		
▸ tubs, replaced		
▸ water heater		
▸ water pipes		

continued on next page

	ORIGINAL COST	ADJUSTED COST
Plumbing — new		
▶ kitchen(s)		
▶ bath(s)		
Heating and A/C		
Electrical — upfitting		
Electrical — new wiring and service		
Insulation		
Drywall or plaster		
Water and sewer main repair		
Well and septic system		
Cabinets, bath vanities		
Interior trim materials		
Interior doors		
Interior trim labor		
Painting, interior		
Painting, exterior		
Appliances/Light fixtures		
Floor coverings		
Drives, walks, patios		
Cleaning		
Gutters, screens, miscellaneous		
Wallpaper		
Hardware and bath accessories		
Landscaping		
Miscellaneous allowance		

Testing for Environmentally Hazardous Materials

If you are planning a major renovation of an older home, tests should be performed to determine the presence of materials which could be hazardous to your family's health. Most dangerous are asbestos, lead, and radon. You may already have test results from when you purchased your house, but if not, make sure you know if construction work will turn up any problems. Your state Department of Environmental Protection can be a good source of information for testing resources.

Asbestos was used for its fireproofing and insulating properties in many houses built between 1920 and 1960, and most often can be found in pipe insulation, siding shingles, and 9-inch square flooring tiles. If it's in good condition and it does not pose a health hazard, no laws or regulations require that it be removed. However, building owners are required to keep asbestos in good repair. If a demolition/renovation activity could cause damage to asbestos-containing material, you need to check local and state regulations pertaining to asbestos abatement or removal.

Removal and proper disposal is an expensive process and is required to be performed by licensed professionals. Asbestos siding and flooring can often be covered with new materials, but care must be taken that the asbestos is not broken and is contained under the new covering.

Lead. Almost all houses built before 1960 used lead-based paint (which is now banned), and lead can also turn up in the water system where lead solder was used at pipe joints or from lead pipes in the water mains. Lead poisoning can be very debilitating, especially for children under the age of 6.

There are a number of lead paint test kits available on the market for do-it-yourselfers, but since lead can actually be released into the environment by a sloppily done test, this is best left to a licensed professional who can also recommend remediation procedures if lead is found. Such procedures include encapsulation (covering with special paint), which is the easiest, to removal, which is the most expensive. The safest method for small children is to remove all lead paint to a point 36 inches off the floor.

Homes built before 1930 are most likely to have lead plumbing. Lead in the water system can be detected during the water test, which should be done to detect other minerals and pollutants as well. The water to be tested should have been standing in the pipes for at least twelve hours before being drawn from the tap. A certified testing lab can be found in the Yellow Pages.

Radon is a naturally occurring radioactive gas that emanates from the soil and is the second leading cause of lung cancer in the United States. Nearly 1 out of 15 homes in this country have elevated radon levels, and it makes no difference if they are old or new, drafty or well sealed, or with or without a basement.

Radon can only be detected by testing, and testing devices can be obtained at retail stores. The short-term testing method requires a minimum of 48 hours for accurate results. Testing should be done at the lowest level of the home that will be used for occupancy. Make sure the testing device is certified by the U.S. Environmental Protection Agency (EPA), which also can provide you with more information on reducing radon levels in the home. Radon reduction can be expensive or not, depending on the design of the home. Radon Reduction Contractors are certified by the EPA or state agencies.

Compared to radon entering the home through the soil, radon entering the home through water will in most cases be a small source of risk. It is usually not a concern in public water supplies, but has been found in well water. If elevated levels have been found in the short-term air test mentioned above, and the water comes from a well, have the water tested.

Tearing Down and Building Up

There are two phases in renovation: tearing down and rebuilding. Both stages have their own set of challenges. Following is the general order of steps for the completion of the tearing down phase of a renovation project.

Wall Removal. This phase follows your master plan and blueprints. If only a few walls are to be removed, your carpenters can usually handle first the removal of the plaster or wallboard for those walls only. If all the plaster is to be removed from all the walls, remove the plaster before tearing out any walls. Your carpenter may want to hold off on wall removal until ready to do any additional framing or bracing. This is fine. Discuss this scheduling, as well as costs, in the planning stage.

Plaster Removal. Other than cutting holes to change wiring or plumbing, I recommend removing plaster only if it is severely cracked or falling down. Plumbing removal may necessitate removing some plaster, but usually in small areas that can easily be patched. Patching is always cheaper — even covering cracked walls with drywall is less expensive than totally gutting them. Keep in mind that all interior trim must be removed, and can seldom be saved, in total gutting.

Wiring and Plumbing Removal. If you are doing a total gut and have removed all the plaster or drywall, then this is the time to remove any plumbing pipes, electrical wiring, radiator pipes, or boilers you have decided to replace.

Once the teardown work is complete, the fun part of the project begins. The following is the general order of steps for completion of the rebuilding phase of a renovation project.

Foundation, Concrete, and Brick Work. Adding or repairing footings and foundations, slab work in basements, and brick work such as chimneys, new foundations, or other brick repairs can be completed before beginning any of the items in this chapter, but I prefer to do all the removal necessary before I start repairing anything.

If your inspection engineer has reported or if you have found evidence that water is or has been in the basement or crawl space, now is the time to call in your waterproofing subcontractor. Both types of space can be waterproofed from the inside, but it is best to do it from the outside. This often requires digging around the foundation and can be expensive, so be sure to get bids. If a footing drain is required, your waterproofing sub is responsible for installing it. Seek additional advice or opinions, if needed, from your architect, inspection engineer, or building inspector.

Rough Carpentry. Your carpenter is one of your key people, for he is needed from the onset of the project for plan review through the initial stages of the renovation. This is the time to remedy all sags in an old house. Sags should always be taken care of prior to doing any new plumbing because correcting them can crack existing pipes.

It is wise to have your carpenter ask the electrician, plumber, and heating/air-conditioning subs what he can do to make their jobs easier. For example, many old houses have receptacles in the baseboards. Removal of the baseboard facilitates replacement of wiring, but the baseboard removal is a job usually handled by your carpenter. The same is true with the removal of kitchen cabinets to aid the plumber. Chases may need to be constructed for pipes, heat vents, A/C lines, and service wires. This is accomplished by furring out a wall, or building a boxlike run from floor to ceiling in closets, or wall-to-wall along the ceiling at an inconspicuous place. These chases or furred areas may or may not be indicated on the plans. If they are not, your subs can work it out on the job.

Roofing. With an existing structure, have any roof repairs done as soon as possible to prevent further deterioration. With additions, the roofing is completed early on for the same reason. Also, with the roof in place, your other subs can work inside on other phases regardless of the weather. Your carpenter will repair any rotted or sagging roof framing and your roofing sub will then install new roofing or patch existing roofing per your specifications.

Note: If there is to be chimney repair or a new chimney, and you are putting on new roofing, have the roofer leave undone an area adjacent to the chimney. This will prevent damage to the new material while the chimney work progresses. Hold back 10 percent of the roofer's contract amount until he can come back and finish this area after the chimney is completed.

Electrical, Plumbing, HVAC. All your electrical, plumbing and HVAC should be completed at this time, prior to insulation of any side walls.

Chimney. Chimneys can now be added or repaired. Old chimneys, if they are structurally sound, should be relined with terra-cotta flue liners and new dampers, if this was deemed necessary when they were inspected. This is faster and less expensive than completely rebuilding.

Exterior Siding and Trim. This phase can be completed while your electrical, plumbing, and heating and air work is being done inside.

Insulation. The cost of energy is sure to keep rising, so it pays to insulate well. Virtually every structure can be insulated without removing the inside walls. This is accomplished by drilling small holes in exterior walls from the outside and blowing fiberglass or cellulose insulation through the holes into the wall-cavity. The holes are then plugged.

The alternative is to remove the interior plaster or drywall of all exterior walls to install insulation batting. This is obviously quite expensive unless you are doing a full gut anyway. Blowing the insulation into the wall cavities provides adequate insulation for most climates, and a reputable firm should do the job neatly. You will not, however, have the vapor barrier that you would get with insulation from inside. You can compensate for this by painting all exterior walls with a vapor barrier paint.

Vapor barriers are important because moisture escapes through the walls, lessening the effectiveness of the insulation and rotting the framing. Before the insulation was added there was no such problem because the moisture went through to the outside. Now it can't, so the object is to keep it inside the heated area. This also adds to your comfort, as moist air is more pleasant than dry air.

Be sure attic areas, crawl spaces, and basements are insulated as well as you can afford. Your utility companies can provide you with information on the amount of insulation needed for lowest rates and greatest savings.

Drywall or Plaster. Now you are ready for the installation or patching of plaster or drywall. Where possible, use drywall to replace missing plaster or to cover cracked plaster. It is far less expensive and looks better. Look for a drywall sub who supplies all materials and removes all trash. Plasterers always supply the necessary materials. In the winter you will need temporary heat to help speed the drywall or plastering job and to prevent cracking caused by freezing — both drywall mud, or filler, and plaster are water-based.

Prime Painting. I find it beneficial to prime the walls as soon as the plaster or drywall sub is finished. Imperfections in the wall show up after priming and serious flaws can be corrected now or later. Discuss this with your drywall or plaster sub beforehand.

Interior Trim. Interior doors, trims, and moldings are repaired, replaced, or added at this time, and any new kitchen or bath cabinets are installed.

Painting. You are now ready for interior as well as exterior painting, although exterior painting could be completed — right after exterior trim. That schedule should be discussed and worked out with your painting sub ahead of time. It is usually easier to complete the whole job at once.

Final Trims. At this time you are ready to install any countertops, vinyl flooring, plumbing fixtures, electrical trims, light fixtures, and final trim for heating and air conditioning — in that order.
Note: If you are finishing or refinishing hardwood floors, have the floors done just prior to laying any carpeting. Install carpeting and hang wallpaper, then proceed with the cleanup. (You can reverse carpeting and cleanup.)

Cleanup. With final cleanup comes trash removal. As general contractor, hiring a dumpster is your responsibility. Some subs will haul away debris from smaller jobs as part of their contract — it never hurts to ask.

Final Inspection and Loan Closing. When all is done, be sure your subs have called for their final inspections and that you have called for a final building inspection. Again, if you don't have a building inspection department, for peace of mind, call in an inspection engineer before final payment to relevant subs. Final inspections will verify compliance with code and that everything works.

When all work has been approved by inspectors, utility companies (if applicable), your lender, and most importantly, you, you are ready to convert any construction financing to the permanent mortgage.

Now you can move in and enjoy.

Starting from Scratch — the Extreme Makeover

If you determine that the effort, cost, and "value added" in renovating a house doesn't make sense or you just want a new house on a particular site and there is a house in the way, then tearing down the existing house and building a new one may be the option for you.

This option usually requires a lot of cash or existing equity in the project. The only legal way to tear down a house is to either own it free and clear or to pay off any existing mortgages. You could get permission from the mortgagor, but this only works if the loan balance is less than the value of the land, since once the house is gone, the only thing of value left is the land! Whatever improvements you put into the house as cash or equity are gone forever once it is torn down.

Here are some other things to think about.

◊ An appraisal will determine what the land alone (with the house torn down) will be worth prior to you getting involved too deeply. Get one!

◊ A construction loan can often pay off an existing mortgage (up to the value of the land). It takes an experienced loan officer to be able to structure an extreme makeover loan (and I speak from that kind of experience).

◊ Other than surmounting the legal and financial obstacles of tearing down a house (and overcoming any emotions you may have about tearing one down), you'll simply be building a new house.

◊ Start at the beginning of this book and proceed with your project. The only additional costs will be demolition and demolition permits, neither of which are cheap.

A NOTE OF CAUTION!

Don't tear down a house that is financed and has a mortgage on it without getting written permission from the lender or paying off the loan first! It's illegal. Your construction lender will pay off existing liens (mortgages) with the first loan disbursement (draw), enabling the teardown.

Building Details

These photos illustrate some helpful tips for handling a few of the details you will encounter in your house construction project.

Waterproofing the Foundation

This foundation is going to be backfilled with dirt or sand before the concrete slab is poured for a three-stall garage. The wall on the right that separates this area from the rest of the basement has been waterproofed to help keep the adjoining basement drier. Waterproofing this type of wall is often overlooked or forgotten until it's too late to do it, so make sure you check.

Managing Waste

Most city locales require one or both of the pictured items on building sites. If they don't, I can only tell you that your subcontractors will use the toilet, but not the dumpster. The only person who will be putting trash in a dumpster is you or someone you hire specifically to clean up the site. What's the alternative to a dumpster? If local regulations permit, have the trash scraped into a pile by a tractor or front-end loader, put in a dump truck, and hauled to the dump several times during construction. The amount of trash created in new construction will astound you.

Fireproofing Details

When your electrician installs recessed lighting, as shown here, be sure that he puts a fireproof baffle between the fixture and the soon-to-be-installed attic insulation. Most building codes require this simple and inexpensive precaution, but it has been overlooked from time to time, resulting in many a house fire. Ready-made baffles are available.

Building a Better Fireplace

Fireplaces, while a pleasure to look at, are usually a horrible waste of energy. Rather than add warmth to a room, they actually pull warm air out of your house while being used. A solution to that dilemma is an energy-efficient fireplace that recirculates heat and also draws the air used for combustion from the outside. If you can't locate this type of unit locally, you can certainly find one on the Internet.

Using Waterproof Drywall

Regular drywall should never be used around a bathtub or shower, even if a vinyl or fiberglass surround will be installed. Here, the drywall has been left out of the tub surround area, and waterproof drywall, or waterproof lathe and cement for ceramic tile, will be used.

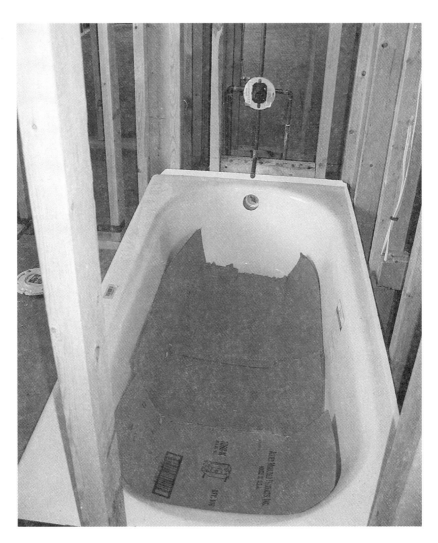

Protecting Your Investment

Bathtubs are installed during the plumbing rough-in. Rough-ins and the construction activity that follows can be rough on finished materials such as bathtubs, whirlpools, and other fixtures. Be sure you or your workers protect such items from scratches and dents by laying down cardboard or wrapping the unit in heavy plastic.

Preventing Leaks

Be sure your plumber protects all plumbing pipes from nails that will be used as construction progresses. The plates at the base of the pipes will protect the plumbing when the drywall and baseboard are nailed up. I have had many a nail cause frustratingly hard-to-find leaks. Most building codes and building inspectors are on top of this, but you should be, too.

Preventing Punctures

The same protection for plumbing pipes is required for electrical wiring. Many a drywall nail has caused a problem or two. Do you see the metal protection plate in this photo? It is protecting the wires that run through the middle of the stud.

Protecting Your Roof

When using brick as a veneer siding, have your roofers leave off the shingles near any area where brick is to be installed later on. This will keep your new roof from being damaged by the bricklayers. The small area left unshingled can be kept covered until the brickwork is completed. Not a perfect system, but it works.

Finishing the Garage

Finish the inside of your garage with drywall. It's inexpensive, it looks good, and adds resale value to your house. In this photo you can see that the wall adjoining the house has been well insulated. Wall insulation has a vapor barrier of either paper or plastic on one side of the insulation. It must always be on the side towards the heated area.

If you are going to heat your garage (and many people do), be sure that you insulate the garage walls as well as the house walls and use an insulated garage door and windows. If you have any questions pertaining to insulation, the best source of answers is your local utility company (gas or electric).

Saving Energy Costs

As the builder, you can choose the most energy-efficient appliances available. The water heater and furnace pictured here are so efficient that the exhaust from combustion is cool enough to allow the use of plastic exhaust vents, which are easier to plan for and deal with during construction. You'll save all around with lower construction costs and energy bills, and you'll help protect the environment. Now that's cool!

Accessing Filters

Make sure the return air filters on a forced air heating system are easy to access, so that it is convenient to change them frequently. Frequent changing of filters saves energy and money. It also promotes a healthier environment inside the house. In this picture, the return air duct that you can see in the rough-in stage will have a filter installed on a filter grill that is very easy to access.

Building a Temporary Desk

During the course of construction, you will frequently want to review the house plans and other documents with your subs. However, there is never a good place to do this in a construction site. This smart general contractor had his carpenters build a convenient meeting spot.

Using Alternative Building Materials

No, your eyes are not deceiving you. This is a concrete house that uses Styrofoam forms that are relatively easy to install. You can research the pros and cons of this type of construction and many other alternative building methods on the Internet. The Internet is a great tool for research and shopping, but when purchasing construction materials online, remember that returning a water heater or a garage door isn't as simple as sending back a sweater!

Sample Contracts & Forms

The following legal instruments are published as examples. Because of varying state laws, these should not be used by you unless such use is approved by an attorney.

Manager's / Supervisor's Construction Contract

1. GENERAL

This contract dated _____ is between _____ (owner) and _____(manager), and provides for supervision of construction by Manager of a residence to be built on Owner's Property at _____, and described as _____. The project is described on drawings dated _____ and specifications dated _____, which documents are a part hereof.

2. SCHEDULE

The project is to start as near as possible to _____, with anticipated completion _____ months from starting date.

3. CONTRACT: FEE AND PAYMENT

    A. Owner agrees to pay Manager a minimum fee of $_____ for the work performed under this contract.

        a. Down payment (due prior to start of work) $_____

        b. Framed $_____

        c. Roof on $_____

        d. Ready for drywall $_____

        e. Trimmed out $_____

        f. Final $_____

B.   Payments billed by Manager are due in full within ten (10) days of bill mailing date.

C.   Final payment to Manager is due in full upon completion of residence; however, Manager may bill upon "substantial completion" (see Paragraph 11 for the definition of terms) the amount of the final payment less 10 percent of the value of work yet outstanding. In such a case, the amount of the fee withheld will be billed upon completion.

4. General Intent of Contract

It is intended that the owner be in effect his own "General Contractor" and that the Manager provide the Owner with expert guidance and advice and supervision and coordination of trades and material delivery. It is agreed that Manager acts in a professional capacity and simply as agent for Owner, and that as such he shall not assume or incur any pecuniary responsibility to contractor, subcontractors, laborers, or material suppliers. Owner will contract directly with subcontractors and obtain from them their certificates of insurance and releases of liens. Similarly, Owner will open his own accounts with material suppliers and be billed and pay directly for materials supplied. Owner shall ensure that insurance is provided to protect all parties of interest. Owner shall pay all expenses incurred in completing the project, except Manager's overhead as specifically exempted in Paragraph 9. In fulfilling his responsibilities to Owner, Manager shall perform at all times in a manner intended to be beneficial to the interests of the Owner.

5. Responsibilities of Manager

General

Manager shall have full responsibility for coordination of trades; ordering materials and scheduling of work; correction of errors and conflicts, if any, in the work, materials, or plans; compliance with applicable codes; judgment as to the adequacy of trades' work to meet standards specified; together with any other function that might reasonably be expected in order to provide Owner with a single source of responsibility for supervision and coordination of work.

Specific Responsibilities
1. Submit to Owner in a timely manner a list of subcontractors and suppliers Manager believes competent to perform the work at competitive prices. Owner may use such recommendations or not at his option.
2. Submit to Owner a list of items requiring Owner's selection, with schedule dates for selection indicated, and recommended sources indicated.
3. Obtain in Owner's name(s) all permits required by governmental authorities.
4. Arrange for all required surveys and site engineering work.
5. Arrange for all the installation of temporary services.
6. Arrange for and supervise clearing and disposal of stumps and brush, and all excavating and grading work.
7. Develop material lists and order all materials in a timely manner from sources designated by Owner.
8. Schedule, coordinate, and supervise the work for all subcontractors designated by Owner.
9. Review, when required by Owner, questionable bills and recommend payment action to Owner.
10. Arrange for common labor for hand digging, grading, and clean-up during construction, and for disposal of construction waste.
11. Supervise the project through completion as defined in Paragraph 11.

6. RESPONSIBILITIES OF OWNER

Owner agrees to:
1. Arrange all financing needed for project so that sufficient funds exist to pay all bills within ten (10) days of their presentation.
2. Select subcontractors and suppliers in a timely manner so as not to delay the work. Establish charge accounts and execute contracts with same, as appropriate, and inform Manager of accounts opened and of Manager's authority in using said accounts.
3. Select items requiring Owner selection and inform Manager of

4. Inform Manager promptly of any changes desired or other matters affecting schedule so that adjustments can be incorporated in the schedule.

5. Appoint an agent to pay for work and make decisions on Owner's behalf in cases where Owner is unavailable to do so.

6. Assume complete responsibility for any theft and vandalism of Owner's property occurring on the job. Authorize replacement/repairs required in a timely manner.

7. Provide a surety bond for his lender if required.

8. Obtain release of liens documentation as required by Owner's lender.

9. Provide insurance coverage as listed in Paragraph 12.

10. Pay promptly for all work done, materials used, and other services and fees generated in the execution of the project, except as specifically exempted in Paragraph 9.

7. EXCLUSIONS

The following items shown on the drawings and/or specifications are NOT included in this contract, insofar as Manager supervision responsibilities are concerned: (List below)

8. EXTRAS/CHANGES

Manager's fee is based on supervising the project as defined in the drawings and specifications. Should additional supervisory work be required because of Extras or Changes occasioned by Owner, unforeseen site conditions, or governmental authorities, Manager will be paid an additional fee of 15 percent of cost of such work. Since the basic contract fee is a *minimum fee*, no downward adjustment will be made if the scope of work is reduced, unless contract is canceled in accordance with Paragraphs 13 or 14.

9. MANAGER'S FACILITIES

Manager will furnish his own transportation and office facilities for Manager's use in supervising the project at no expense to Owner. Manager shall provide general liability and workmen's compensation insurance coverage for Manager's direct employees only at no cost to Owner.

10. Use of Manager's Accounts

Managers may have certain "trade" accounts not available to Owner that Owner may find it to his advantage to utilize. If Manager is billed and pays such accounts from Manager's resources, Owner will reimburse Manager within ten (10) days of receipt of Manager's bill at cost plus 8 percent of such materials/services.

11. Project Completion

A. The project shall be deemed completed when all the terms of this contract have been fulfilled and a Residential Use Permit has been issued.

B. The project shall be deemed "substantially complete" when a Residential Use Permit has been issued and less than Five Hundred Dollars ($500) of work remains to be done.

12. Insurance

Owner shall ensure that workmen's compensation and general liability insurance are provided to protect all parties of interest and shall hold Manager harmless from all claims by subcontractors, suppliers and their personnel, and for personnel arranged for by Manager on Owner's behalf, if any.

Owner shall maintain fire and extended coverage insurance sufficient to provide 100 percent coverage of project value at all stages of construction, and Manager shall be named in the policy to insure his interest in the project.

Should Owner or Manager determine that certain subcontractors, laborers, or suppliers are not adequately covered by general liability or workmen's compensation insurance to protect Owner's and/or Manager's interests, Manager may, as agent of Owner, cover said personnel on Manager's policies, and Owner shall reimburse Manager for the premium at cost plus 10 percent.

13. Manager's Right to Terminate Contract

Should the work be stopped by any public authority for a period of thirty (30) days or more through no fault of the Manager, or should work be stopped through act or neglect of Owner for ten (10) days or more, or should Owner fail to pay Manager any payment due within ten (10) days written notice to Owner, Manager may stop work and/or terminate this contract and recover from Owner payment for all work completed as a proration of the total contract sum, plus 25 percent of the fee remaining to be paid if the contract were completed as liquidated damages.

14. OWNER'S RIGHT TO TERMINATE CONTRACT

Should the work be stopped or wrongly prosecuted through act or neglect of Manager for ten (10) days or more, Owner may so notify Manager in writing. If work is not properly resumed within ten (10) days of such notice, Owner may terminate this contract. Upon termination, entire balance then due Manager for that percentage of work then completed, as a proration of the total contract sum, shall be due and payable and all further liabilities of Manager under this contract shall cease. Balance due to Manager shall take into account any additional cost to Owner to complete the house occasioned by manager.

15. MANAGER/OWNER'S LIABILITY FOR COLLECTION EXPENSES

Should Manager or Owner respectively be required to collect funds rightfully due him through legal proceedings, Manager or Owner respectively agrees to pay all costs and reasonable attorney's fees.

16. WARRANTIES AND SERVICE

Manager warrants that he will supervise the construction in accordance with the terms of this contract. No other warranty by manager is implied or exists.

Subcontractors normally warrant their work for one year, and some manufacturers supply yearly warranties on certain of their equipment; such warranties shall run to the Owner and the enforcement of these warranties is in all cases the responsibility of the Owner and not the Manager.

(Manager) _____ (seal) Date:

(Owner) _____ (seal) Date:

(Owner) _____ (seal) Date:

Contract to Build House
(Cost Plus Fee)

Contractor: _____

Owner: _____ Date: _____

Owner is or shall become fee simple owner of a tract or parcel of land known or described as:_____.

Contractor hereby agrees to construct a residence on the above described lot according to the plans and specification identified as: Exhibit A — plans and specifications drawn _____ by _____.

Owner shall pay Contractor for the construction of said house cost of construction and a fee of _____. Cost is estimated in Exhibit B. Each item in Exhibit B is an estimate and is not to be construed as an exact cost.

Owner shall secure/has secured financing for the construction of said house in the amount of cost plus fee, which shall be disbursed by a savings and loan or bank from time to time as construction progresses, subject to a holdback of no more than 10 percent. Owner hereby authorizes Contractor to submit a request for draws in the name of Owner under such loan up to the percentage completion of construction and to accept said draws in partial payment hereof. In addition, it is understood that the Contractor's fee shall be paid in installments by the savings and loan or bank at the time of and as a part of each construction draw as a percentage of completion, so that the entire fee shall be paid at or before the final construction draw.

Contractor shall commence construction as soon as feasible after closing of the construction loan and shall pursue work to a scheduled completion on or before seven months from commencement, except if such completion shall be delayed by unusually unfavorable weather, strikes, natural disasters, unavailability of labor or materials, or changes in the plans or specifications.

Contractor shall build the residence in substantial compliance with the plans and specifications and in a good and workmanlike manner and shall meet all building codes. Contractor shall not be responsible for failure of materials or equipment not Contractor's fault. Except as herein set out, Contractor shall make no representations or warranties with respect to the work to be done hereunder.

Owner shall not occupy the residence and Contractor shall hold the keys until all work has been completed and all monies due Contractor hereunder shall have been paid.

Owner shall not make changes to the plans or specifications until such changes shall be evidenced in writing; the costs, if any, of such changes shall be set out; and the construction lender and Contractor shall have approved such changes. Any additional costs thereof shall be paid in advance, or payment guaranteed in advance for the work being accomplished.

Contractor shall not be obligated to continue work hereunder in the event Owner shall breach any term or condition hereof, or if for any reason the construction lender shall cease making advances under the construction loan upon proper request thereof.

Any additional or special stipulations attached hereto and signed by the parties shall be and are made a part thereof.

Contractor: _____ (seal)

Owner: _____ (seal)

             _____ (seal)

Contract to Build House
(Contract Bid)

Contractor: _____

Owner: _____   Date: _____

Owner is or shall become fee simple owner of a tract or parcel of land known or described as _____

Contractor hereby agrees to construct a residence on the above described lot according to the plans drawn by _____, and the specifications herein attached.

Owner shall pay Contractor for the construction of said house $_____.

Prior to commencement hereunder, Owner shall secure financing for the construction of said house in the amount of $_____, which loan shall be disbursed from time to time as construction progresses, subject to a holdback of no more than 10 percent. Owner hereby authorizes Contractor to submit a request for draws in the name of the Owner from the savings and loan, or similar institution, up to the percentage completion of construction and to accept said draws in partial payment thereof.

Contractor shall commence construction as soon as feasible after closing and shall pursue work to a scheduled completion on or before seven months from commencement, except if such completion shall be delayed by unusually unfavorable weather, strikes, natural disasters, unavailability of labor or materials, or changes in the plans and specifications.

Contractor shall build the residence in substantial compliance with the plans and specifications and in a good and workmanlike manner and shall meet all building code requirements. Contractor shall not be responsible for failure of materials or equipment not Contractor's fault. Except as herein set out, Contractor shall make no representations or warranties with respect to the work to be done hereunder.

Owner shall not occupy the residence and Contractor shall hold the keys until all work has been completed and all monies due Contractor hereunder shall have been paid.

Owner shall not make any changes to the plans and specifications until such changes shall be evidenced in writing; the costs, if any, of such changes shall be set out; and any additional costs thereof shall be paid in advance of the work being accomplished.

Contractor shall not be obligated to continue work hereunder in the event Owner shall breach any term or condition hereof, or if for any reason the construction draws shall cease to be advanced upon proper request thereof.

Any additional or special stipulations attached hereto and signed by the parties shall be and are made a part hereof.

Contractor: _____ (seal)

Owner: _____ (seal)

_____ (seal)

Description of Materials

❏ Proposed Construction   No._____

❏ Under Construction

Property address_____

City_____ State _____

Mortgagor or Sponsor_____

Contractor or Builder _____

INSTRUCTIONS

1. For additional information on how this form is to be submitted, etc., see the instructions applicable to the FHA Application for Mortgage Insurance or VA Request for Determination of Reasonable Value, as the case may be.

2. Describe all materials and equipment to be used, whether or not shown on the drawings, by marking an X in each appropriate check-box and entering the information called for in each space. If space is inadequate, enter "See misc." and describe under item 27 or on an attached sheet. The use of paint containing more than one percent lead by weight is prohibited.

3. Work not specifically described or shown will not be considered unless required, then the minimum acceptable will be assumed. Work exceeding minimum requirements cannot be considered unless specifically described.

4. Include no alternates, or "equal" phrases, or contradictory items. (Consideration of a request for acceptance of substitute materials or equipment is not thereby precluded.)

5. Include signatures required at the end of this form.

6. The construction shall be completed in compliance with the related drawings and specifications, as amended during processing. The specifications include this Description of Materials and the applicable Minimum Property Standards.

1. Excavation
Bearing soil, type _____

2. Foundations
Footings: concrete mix _____; strength psi   Reinforcing _____

Foundation wall: material _____   Reinforcing _____

Interior foundation wall: material _____ Party foundation wall _____

Columns: material and sizes _____ Piers: material and reinforcing _____

Girders: Material and sizes_____ Sills: material _____

Basement entrance areaway_____ Window areaways _____

Waterproofing _____

Footing drains_____

Termite protection_____

Basementless space: ground cover_____; insulation _____;

      foundation vents_____

Special foundations _____

Additional information _____

_____

3. Chimneys
Material_____   Prefabricated _____
                                              (make and size)

Flue lining: material_____ Heater flue size_____

Fireplace flue size _____

Vents *(material and size):* gas or oil heater _____ ;

      water heater _____

Additional information_____

_____

4. Fireplaces
Type ❏ solid fuel; ❏ gas-burning; ❏ circulator _____
                                                 (name and size)

Ash dump and clean-out_____ Fireplace: facing _____ ;

      lining _____ ; hearth _____;

      mantel _____

Additional information_____

_____

5. EXTERIOR WALLS

Wood frame: wood, grade, and species _____   ❑ Corner bracing

Building paper or felt _____ Sheathing _____;

    thickness_____; width _____;

    ❑ solid;   ❑ spaced _____" o.c.;   ❑ diagonal: _____

Siding _____; grade _____;

    type _____; size _____; exposure_____"; fastening _____

Shingles _____; grade_____;

    type _____; size _____;

    exposure _____; fastening _____

Stucco _____; thickness _____";

Lath _____ weight _____lbs.

Masonry veneer _____ Sills _____ Lintels _____ Base flashing _____

Masonry: ❑ solid ❑ faced ❑ stuccoed; total wall thickness _____";

    facing thickness_____; facing material _____

    Backup material _____; thickness ____"; bonding _____

    Door sills _____Window sills _____ Lintels _____ Base flashing ____

    Interior surfaces: damp proofing, _____ coats of _____; furring _____

Additional information_____

Exterior painting: material _____; number of coats _____

Gable wall construction: ❑ same as main walls ❑ other construction

6. FLOOR FRAMING

Joists: wood, grade, and species _____; other _____;

    bridging _____; anchors _____

Concrete slab: ❑ basement floor ❑ first floor ❑ ground supported

    ❑ self-supporting mix_____; thickness _____";

    reinforcing _____ insulation _____; membrane _____

Fill under slab: material _____ thickness _____"

Additional information_____

7. SUBFLOORING *(describe underflooring for special floors under item 21)*

Material: grade and species _____; size _____; type _____

Laid: ❑ first floor ❑ second floor ❑ attic ____ sq. ft. ❑ diagonal

    ❑ right angles

Additional information _____

8. FINISH FLOORING *(Wood only. Describe other finish flooring under item 21)*

location	rooms	grade	species	thickness	width	bldg.	paper	finish

First floor _____

Second floor _____

Attic floor _____ sq. ft. _____

Additional information _____

9. PARTITION FRAMING

Studs: wood, grade, and species _____; size and spacing _____; other _____

Additional information _____

10. CEILING FRAMING

Joists: wood, grade, and species _____; other _____; bridging _____

Additional information _____

11. ROOF FRAMING

Rafters: wood, grade, and species _____

Roof trusses (see detail): grade and species _____

Additional information _____

12. ROOFING

Sheathing: wood, grade, and species _____ ❏ solid ❏ spaced ___" o.c.

Roofing _____; grade _____; size _____; type _____;

Underlay _____; weight or thickness_____; size _____; fastening _____

Built-up roofing _____; number of plies _____; surfacing material_____

Flashing: material ___; gauge or weight___; ❏ gravel stops ❏ snow guards

Additional information _____

13. GUTTERS AND DOWNSPOUTS

Gutters: material _____; gauge or weight _____; size _____; shape _____

Downspouts: material _____; gauge or weight _____; size _____;

    shape _____; number _____

Downspouts connected to ❏ storm sewer ❏ sanitary sewer ❏ dry well

    ❏ splash blocks: material and size _____

Additional information _____

14. LATH AND PLASTER

Lath ❏ walls ❏ ceilings: material _____ ; weight or thickness _____

Plaster: coats _____ ; finish _____

Drywall ❏ walls ❏ ceilings: material _____ ; thickness _____ ; finish _____

Joint treatment _____

15. DECORATING *(paint, wallpaper, etc.)*

rooms	wall finish material and application	ceiling finish material and application

Kitchen _____

Bath _____

Other _____

Additional information _____

16. INTERIOR DOORS AND TRIM

Doors: type _____ ; material _____ ; thickness _____

Door trim: type _____ ; material _____ ;

Base: type _____ ; material _____ ; size _____

Finish: doors _____ ; trim _____

Other trim *(item, type, and location)* _____

Additional information _____

17. WINDOWS

Windows: type _____ ; make _____ ; material _____ ; sash thickness ____

Glass: grade _____ ; ❏ sash weights ❏ balances, type _____ ;

head flashing _____

Trim: type _____ ; material _____

Paint _____ ; number coats _____

Weatherstripping: type_____ ; material _____ ;

Storm sash, number _____

Screens: ❏ full ❏ half; type _____ ; number _____ ;

screen cloth material _____

Basement windows: type _____ ; material _____ ; screens, number ___ ;

Storm sash, number_____

Special windows _____

Additional information _____

18. Entrances and Exterior Detail

Main entrance door: material _____; width _____; thickness ____";

Frame: material_____; thickness ____ "

Other entrance doors: material _____; width _____; thickness _____"

Frame: material _____ thickness _____"

Head flashing _____ Weatherstripping: type _____; saddles _____

Screen doors: thickness ____"; number _____ screen cloth material

Storm doors: thickness _____"; number _____

Combination storm and screen doors; thickness ____"; number _____;

    screen cloth material _____

Shutters: ❑ hinged ❑ fixed Railings _____; Attic louvers_____

Exterior millwork; grade and species _____

Paint _____; number coats _____

Additional information _____

19. Cabinets and Interior Detail

Kitchen cabinets, wall units: material _____; lineal feet of shelves ____;

    shelf width ____

    Base units: material _____; countertop _____; edging _____

    Back and end splash ____ Finish of cabinets ____; number coats ____

Medicine cabinets: make _____; model _____

Other cabinets and built-in furniture _____

Additional information _____

20. Stairs

	Treads material thickness	Risers material thickness	Strings material size	Handrail material size	Balusters material size
Basement					
Main					
Attic					

Disappearing: make and model number _____

Additional information _____

21. SPECIAL FLOORS AND WAINSCOT

		Wainscot	Floors
location	material, color, border, sizes, gauge.	threshold mat'l wall base mat'l	underfloor mat'l

Kitchen_____

Bath _____

location	material, color, border, cap, sizes, gauge	height	height over tub	height in showers (from floor)

Bath _____

_____

Bathroom accessories: ❏ Recessed: material _____; number_____

❏ Attached: material _____; number _____

Additional material _____

22. PLUMBING

fixture	number location	make	mfr.'s fixture identification no.	size	color

Sink _____

Lavatory _____

Water closet _____

Bathtub _____

❏ Shower over tub _____

❏ Stall shower _____

Laundry trays _____

❏ Curtain rod   ❏ Door   ❏ Shower pan: material _____

Water supply: ❏ public ❏ community system ❏ individual (private) system*

Sewage disposal: ❏ public ❏ community system ❏ individual (private) system*

Show and describe individual system in complete detail in separate drawings and specifications according to requirements.

House drain (inside) ❏ cast iron ❏ tile ❏ other _____

House sewer (outside) ❏ cast iron ❏ tile ❏ other _____

Water piping ❏ galvanized steel ❏ copper tubing ❏ other _____

Sill cocks, number _____

Domestic water heater: type _____; make and model _____;

Heating capacity _____ gph. 100° rise.

Storage tank: material _____; capacity _____ gallons

Gas service ❏ utility company ❏ liq. pet. gas ❏ other_____

Gas piping ❏ cooking ❏ house heating

Footing drains connected to ❏ storm sewer ❏ sanitary sewer ❏ dry well

Sump pump: make and model _____; capacity _____; discharges into _____

23. HEATING

❏ hot water ❏ steam ❏ vapor ❏ one-pipe system ❏ two-pipe system

❏ radiators ❏ convectors ❏ baseboard radiation

    make and model _____

Radiant panel: ❏ floor ❏ wall ❏ ceiling Panel coil: material _____

❏ circulator ❏ return pump Make and model_____; capacity ____ gpm.

Boiler: make and model _____

    output _____ Btuh; net rating _____ Btuh

Additional information_____

Warm air: ❏ gravity ❏ forced   type of system _____

    Duct material: supply _____; return _____

    Insulation _____ thickness _____ ❏ outside air intake

    Furnace: make and model _____

    Input _____ Btuh; output _____ Btuh

Additional information _____

❏ Space heater   ❏ floor furnace   ❏ wall heater   input ____ Btuh;

    output _____ Btuh; number units _____

    make, model _____

Additional information _____

Controls: make and types _____

Additional information _____

Fuel: ❏ Coal ❏ oil ❏ gas ❏ liq. pet. gas ❏ electric ❏ other_____;

    storage capacity _____

Additional information _____

Firing equipment furnished separately: ❏ Gas burner, conversion type_____

❏ Stoker: hopper feed _____ ❏ bin feed

    oil burner: ❏ pressure atomizing _____ ❏ vaporizing _____

    make and model _____ control _____

Additional information _____

Electric heating system: type _____

    input _____ watts; @ _____ volts; output _____ Btuh

Additional information_____

Ventilating equipment: attic fan, make and model _____

    capacity _____ cfm _____

    kitchen exhaust fan, make and model _____

Other heating, ventilating, or cooling equipment _____

24. ELECTRIC WIRING

Service: ❏ overhead ❏ underground_____ Panel: ❏ fuse box ❏ circuit breaker

make _____ AMPs _____ no. circuits _____

Wiring: ❏ conduit ❏ armored cable ❏ nonmetallic cable ❏ knob and tube

❏ other _____

Special outlets: ❏ range ❏ water heater ❏ other _____

❏ Doorbell ❏ chimes ❏ push-button locations _____

Additional information _____

25. LIGHTING FIXTURES

Total number of fixtures _____

Total allowance for fixtures, typical installation, $_____

Nontypical installation _____

Additional information _____

26. INSULATION

location	thickness	material, type, and method of installation	vapor barrier
Roof			
Ceiling			
Wall			
Floor			

Additional information _____

Hardware *(make, material, and finish):* _____

_____

_____

Special Equipment *(State material or make, model, and quantity. Include only equipment and appliances that are acceptable by local law, custom and applicable FHA standards. Do not include items that, by established custom, are supplied by occupant and removed when he vacates premises or chattels prohibited by law from becoming realty.)*

_____

_____

_____

_____

_____

_____

27. Miscellaneous

(Describe any main dwelling materials, equipment, or construction items not shown elsewhere; or use to provide additional information where the space provided was inadequate. Always reference by item number to correspond to numbering used on this form.)

_____

_____

_____

_____

_____

_____

Porches_____

_____

_____

Terraces_____

_____

_____

Garages_____

_____

_____

Walks and Driveways

Driveway: width _____; base material _____"; thickness _____"

Surfacing material _____; thickness _____"

Front walk: width _____; material _____; thickness _____"

Service walk: width _____; material _____; thickness _____"

Steps: material _____; treads_____"; risers _____"

Cheek walls _____

Other On-site Improvements

(Specify all exterior on-site improvements not described elsewhere, including items such as unusual grading, drainage structures, retaining walls, fences, railings, and accessory structures.)

_____

_____

Landscaping, Planting, and Finish Grading

Topsoil _____" thick: ❏ front yard   ❏ side yards

    ❏ rear yard to _____ feet behind main building

Lawns (seeded, sodded, or sprigged): ❏ front yard _____;

    ❏ side yards _____; ❏ rear yard _____

Planting: ❏ as specified and shown on drawings ❏ as follows:

    Shade trees, deciduous, _____ " caliper

    Low-growing trees, deciduous _____ 'to _____ '

    High-growing shrubs, deciduous _____ ' to _____ '

    Medium-growing shrubs, deciduous _____ ' to _____ '

    Low-growing shrubs, deciduous _____ ' to _____ '

    Evergreen trees _____ ' to _____ ', B&B

    Evergreen shrubs _____ ' to _____ ', B&B

    Vines, 2-year _____

Identification: This exhibit shall be identified by the signature of the builder, or sponsor, and/or the proposed mortgagor if the latter is known at the time of application.

Date _____ Signature _____

    _____ Signature _____

FHA Form 2005
VA Form 26-1852

Certificate of Insurance

ALLSTATE INSURANCE COMPANY HOME OFFICE — NORTHBROOK, ILLINOIS

Name and Address of Party to
Whom this Certificate is Issued

Name and Address of Insured

INSURANCE
IN FORCE

Types of Insurance and Hazards	Policy Forms	Limits of Liability	Policy Number	Expiration Date
Workmen's Comp.		STATUTORY*		
Employers' Liability	**Standard**	$_____ PER ACCIDENT (Employer's Liability only) *Applies only in following state(s):		

Automobile Liability		Bodily Injury Each	Property Damage		
❑ owned only	❑ Basic	$ PERSON			
❑ non-owned only	❑ Comprehensive	$ ACCIDENT $			
❑ hired only	❑ Garage	$ OCCURRENCE $			
		Bodily Inj. and Prop. Dam. (single limit)			
❑ owned, non-owned, and hired		$ EACH ACCIDENT			
		$ EACH OCCURRENCE			

General Liability		Bodily Injury Property Damage	
❑ Premises — O.L.&T.	❑ Schedule	$ EACH PERSON	
❑ Operations — M.&C.		$ EACH $ ACCIDENT	
❑ Elevator	❑ Comprehensive	$ EACH $ OCCURRENCE	
		AGGREG.	
❑ Products/Completed Operations		$ PROD. $ COMP. OPTNS.	
		AGGREGATE OPERATIONS $	
❑ Protective (Independent Contractors)	❑ Special Multi-Peril	AGGREGATE PROTECTIVE $	
		AGGREGATE CONTRACTUAL $	
❑ Endorsed to cover contract between insured and		Bodily Inj. and Prop. Dam. (single limit)	
_____		$ EACH ACCIDENT	
		$ EACH OCCURRENCE	
date _____		$ AGGREGATE	

The policies identified above by number are in force on the date indicated below. With respect to a number entered under policy number, the type of insurance shown at its left is in force, but only with respect to such of the hazards, and under such policy forms, for which an "X" is entered, subject, however, to all the terms of the policy having reference thereto. The limits of liability for such insurance are only as shown above. This Certificate of Insurance neither affirmatively nor negatively amends, extends, nor alters the coverage afforded by the policy or policies numbered in this Certificate.

In the event of reduction of coverage or cancellation of said policies, the Allstate Insurance Company will make all reasonable effort to send notices of such reduction or cancellation to the certificate holder at the address shown above.

This certificate is issued as a matter of information only and confers no rights upon the certificate holder.

Date _____ , 20_____ By _____
Authorized Representative

U454-16
(8-81)

APPENDIX 3

Reading Plans

It's a good idea to study a number of different house plans before settling on the plan that is best for you and your family. You can also
spend a few weekends visiting houses in your area that are currently
under construction or attending open houses by realtors. The more
houses you see, the more ideas you'll find for your own home. Look
for designs that work well for you (a den instead of a formal dining
room? A mudroom or pantry next to the kitchen? Perhaps a second-
storey porch off the master bedroom?) and ones that don't (you may
not want a playroom in the basement or huge walk-in closets). As you
develop your ideas for the best plan for your new house, you can mix
and match and create the perfect space.

Once you have your personal house plans in front of you, study
them carefully and become thoroughly familiar with all the details
before construction begins. You will consult these plans often during
the building process. A thorough review of them should give you the
feel of actually living in the house. Think about traffic patterns, common spaces, entrances and exits, and where your family will spend
most of its time. How will you arrange all your furniture? Is there an
unused spot that might accommodate some built-in shelves or a corner cabinet? Where will you put the TV and all your books?

The drawings on pages 140–149 are examples of typical plans.
Take a moment to look them over.

The **foundation plan** shows all of the concrete work associated with this house, including the walls, piers, foundation for fireplace, and slabs for the garage and the patio. The furnace and hot-water heater could be placed here as well, or else in a crawl space or under stairs on the first floor.

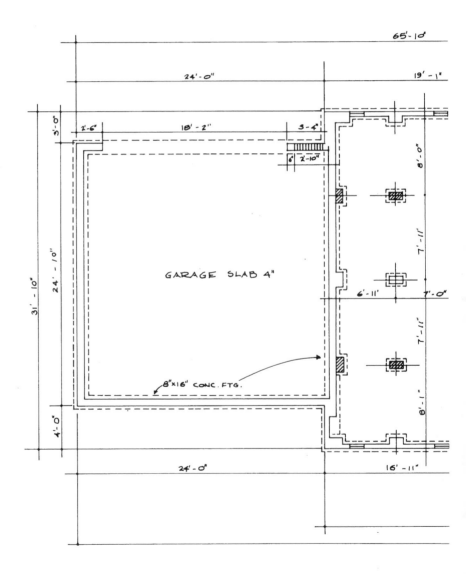

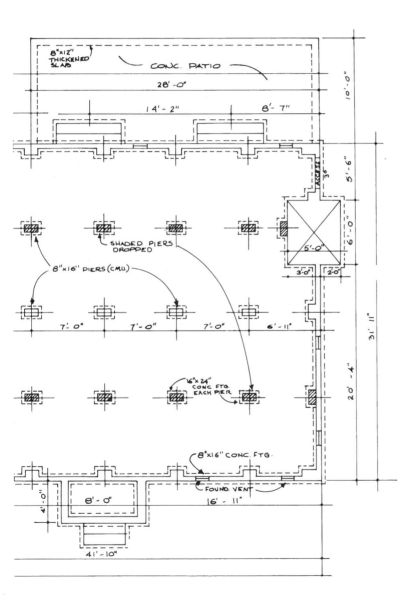

Study your **floor plan** and ask yourself specific questions about living in a house that is built to your plan. Will you be happy with only a sliding glass door in the family room? Do you want both a formal dining room and a separate dinette, or would you prefer to make

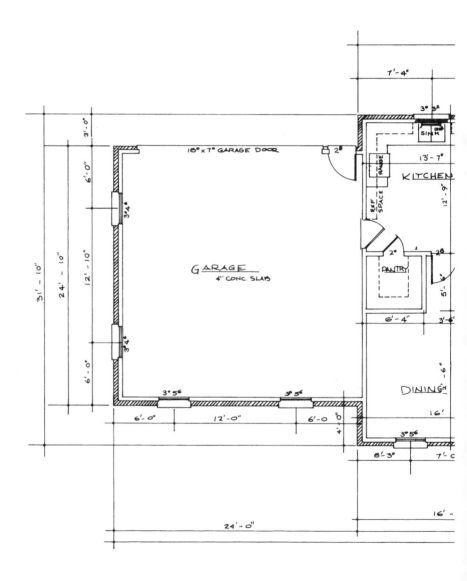

other use of one of those spaces? Do you want your garage doors opening to the rear of the house, or would you prefer parking closer to the front entrance?

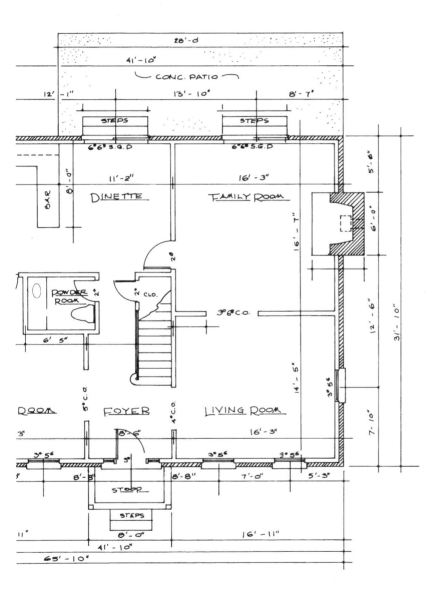

This **detail sheet** shows the second floor. Have you always wanted a fireplace in your master bedroom? It probably could be arranged — at extra cost. Will you be happy with the laundry up there (where most dirty clothes end up) or would you prefer it on the first floor?

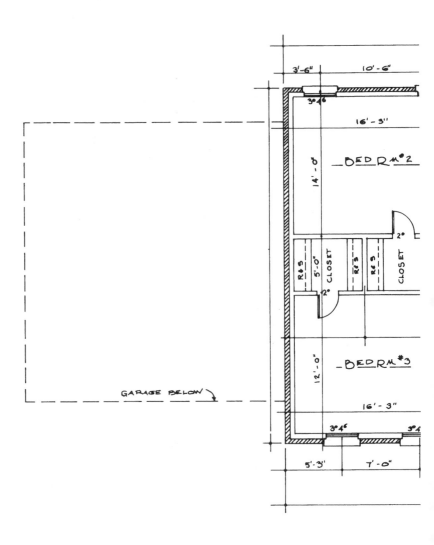

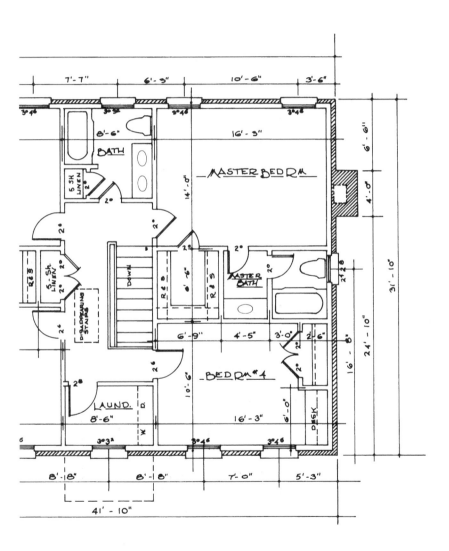

Study your **detail sheets** carefully. Is the kitchen layout what you want, or are there parts of it that will become irritations when it is used?

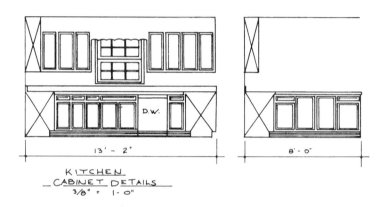

KITCHEN
CABINET DETAILS
3/8" = 1-0"

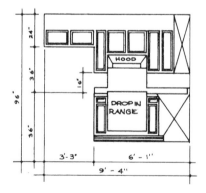

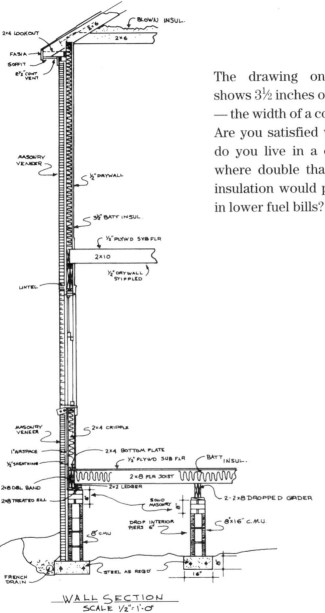

The drawing on this page shows $3\frac{1}{2}$ inches of **insulation** — the width of a common 2 x 4. Are you satisfied with that, or do you live in a cold climate where double that amount of insulation would pay for itself in lower fuel bills?

And finally, the **outside elevations.** Are you happy with their appearance? Do you want that outside chimney, or would you prefer to see it inside, where its stored heat would be fed back into the

Right Side

Rear Elevation

home? Will you feel proud of this house when you drive up to it — or when your friends do?

Left Side

Front Elevation

Metric Conversion Chart

WHEN THE MEASUREMENT GIVEN IS	TO CONVERT IT TO	MULTIPLY IT BY
inches	centimeters	2.54
feet	meters	0.305
miles	kilometers	1.6
square feet	square meters	0.0929
cubic yards	cubic meters	0.765
pounds	kilograms	0.45
°F	°C	°F – 32 x ⁵⁄₉

Glossary

Adjustable Rate Mortgage (ARM). A mortgage in which the interest rate is adjusted periodically up or down, usually once or twice a year.

Annual Percentage Rate (APR). The total amount of your mortgage loan package (interest, loan fees, points, or other charges) expressed as a percentage of the loan amount. The APR is usually slightly higher than the actual interest rate.

Appraised Value. An estimate of the value of property, such as a house.

Balloon (Payment) Mortgage. Usually a short-term fixed-rate loan, it involves small payments for a certain period of time and one large payment for the remaining amount of the principal at a time specified in the contract.

Binder. Money given by a buyer to a seller as part of the purchase price to bind a transaction or ensure payment (also called earnest money).

Breaker. Electrical circuit breaker. The modern version of the old-time fuse.

Brick Veneer. Brick used in lieu of siding.

Bridge Loan. A short-term loan to bridge the time between the purchase of one house and the sale of another.

Bridging. Small pieces of wood or metal used to brace floor joists.

Carpet Underlayment. Either plywood or pressed wood (particleboard) placed over the subfloor to make the floor more solid.

Chase. A channel built into a wall or ceiling to hold wiring, plumbing, or vents.

Clear Title. A title (proof of ownership) of any property (land, auto, house) that is free of liens, mortgages, judgments, or any other encumbrance.

Closing. The meeting between the buyer, seller, and lender or their agents where the property legally changes hands.

Closing Costs. Costs the buyer must pay at the time of closing in addition to the down payment, including points, mortgage insurance premium, homeowner's insurance, prepayments for property taxes, and so on. Closing costs usually average 2 to 4 percent of the loan amount.

Color Run. Materials produced using the same batch of dye, such as bricks, carpet, or paint. Subsequent batches may vary in color.

Contingencies. Unforeseen or unpredictable additional costs.

Contract Price. A pre-agreed set price for a service or product.

Course. One row of bricks.

Crawl Space. An area under a floor or roof where there is insufficient space to stand up.

Deed Restrictions. Encumbrances placed on a piece of real estate and filed as a matter of record that restrict or prevent certain uses of the property.

Down Payment. Money paid up front to make up the difference between the purchase price and the mortgage amount.

Draw. A disbursement of money representing a portion or percentage of the entire amount due.

Drip Cap. A protective molding, usually metal, to divert water over an exterior surface.

Dry-in. The stage in construction when walls, roof, and windows are in place so that the interior of the building is protected from rain and snow.

Drywall. Used to finish interior walls; Sheetrock is a commonly used brand.

Elevations. Drawings of the exterior of a structure.

Equity. The difference between what is owed on a property and what it is worth. It equals the percentage of the mortgage that has been paid off, and can be used as collateral for a home equity loan.

Escrow. An account held by the lender into which the homeowner pays money for tax or insurance payments and also for binders to be held pending loan closing.

Expansion Joint. A joint used to separate blocks or units of concrete to prevent cracking due to expansion as a result of temperature changes.

Fascia Boards. The flat, horizontal boards at the edge of the roof.

Flashing. Sheet metal used in waterproofing roof valleys or the seam between roof and chimney.

Flue. A channel for the passage of hot gases and smoke.

Footing. A concrete structure below the frost line supporting the foundation or piers of a house.

Frame Construction. Using wood, as opposed to brick, block, concrete, or steel, for building walls, ceilings, and roofs.

Framing. The construction of the skeleton of a house, including walls, floors, ceiling, and roof.

French Drains. Pipes placed around or beneath a structure to provide positive drainage.

Frost Line. The depth in the earth at which the warmth of the earth prevents freezing or the formation of frost.

Furr out. To make a wall or ceiling deeper in order to provide a chase.

Heat Pump. An electrical appliance for heating and cooling a building. In the heating phase, it draws heat from the outside air (even cold air) and transfers it to the inside of a house or other building. Acts as an air-conditioner in the summer.

HVAC. The heating, ventilating, air-conditioning system of a building.

Interim Financing. A short-term loan that is generally converted to a long-term loan at a later date.

Joists. Small horizontal timbers that support a floor or ceiling.

Load-Bearing Capability. The amount of weight a particular substance can withstand without breaking or bending beyond its design. Used to describe soil, steel, and wood.

Lot Subordination. A process of buying land by which the owner takes a note in lieu of payment and legally takes an interest in that land secondary to a party with primary interest such as a savings and loan.

Manager's Contract. A contract with a general contractor by which the latter agrees to act as a manager to construct your house. Under such a contract, you are able to remain the primary general contractor.

Market Value. The highest price that a buyer would pay and the lowest price a seller would accept on a property. Market value may be different from the price a property could actually be sold for at a given time.

Molding. The interior trim around windows, doors, ceilings, and other features in a house.

Mortar. A mixture of cement, sand, and water used between bricks or concrete blocks to hold them together.

Mortgage Insurance. Money paid to insure the mortgage when the down payment is less than 20 percent.

Note. A written acknowledgment of a debt, such as a promissory note.

O.C. On center. Before lumber widths were standardized, carpenters measured from the center of one stud (or rafter or joist) to the center of the next. Still used to mean inches apart.

Origination Fee. The fee charged by a lender to prepare loan documents, make credit checks, inspect, and sometimes appraise property; usually computed as a percentage of the face value of the loan.

Panel Box. Electrical panel. A metal box containing circuit breakers.

Percolation (perc) Test. Tests to determine if the soil on a piece of land can support a septic system.

Pier. A vertical structural support. Usually a masonry or metal column that supports the house, porch, or deck.

Pitch (of roof). How steeply the roof rises per foot of horizontal travel. An average pitch is 6/12, or 6 inches of rise per foot.

Pointing up. Touching up minor imperfections and damage in new walls after they have been primed but before final painting occurs.

Points (Loan Discount Points). Prepaid interest assessed at closing by the lender. Each point is equal to 1 percent of the loan amount (e.g., two points on a $100,000 mortgage cost $2,000). The purpose of paying points is to obtain a lower interest rate on the mortgage.

Polyethylene (poly). Plastic film used in home construction to provide a moisture barrier.

Prefab Fireplace. A fireplace that is made of metal as opposed to solid masonry. It usually has a metal flue.

Prewire. Wiring for various items (such as lighting, telephones, intercoms, and burglar alarms) that is installed before drywall or paneling is applied to walls.

Quote. A guaranteed price in advance.

Recording Fee. The fee charged to record legal documents in a place of permanent records, such as a county courthouse.

Reinforcing Rod. A steel rod placed in concrete to increase the strength of the concrete. Also known as rebar.

Roof Pitch. The slope of the roof.

Rough-in. The installation of wiring, plumbing, or heat ducts in the walls, floors, or ceilings before those walls, floors, or ceilings are covered with drywall, plaster, or paneling.

Saw Service. Temporary electrical service used during construction.

Second Mortgage. The pledging of property to a lender as security for repayment, using property that has already been pledged for a loan.

Septic System. A means of disposing of sewage in the ground.

Sheathing. Waterproof material placed between the inner and outer walls of a building. Roof sheathing is a layer of plywood to which the roofing material can be attached.

Shim. A thin wedge of wood or metal used to fill in a space.

Sill (sill plate). A horizontal framing member, immediately adjacent to the foundation, that supports a wall or floor.

Slab. A poured concrete foundation for a garage or other structure with no basement.

Soffit. The underside of a roof overhang.

Soffit Vent. A vent in the soffit that allows air circulation under the roofing and between the roof rafters to prevent heat buildup and rotting of the wood elements in the roof.

Specifications (specs). A listing of the particulars of all materials to be used in building or renovating a structure.

Square. The amount of roofing material needed to cover 100 square feet of roof.

Staking. Placing stakes in the ground prior to building to show the location of the corners of the house.

Stud. An upright board in the framing of a wall, usually spaced on 16-inch or 24-inch centers.

Takeoff. The compilation of a list of materials used for a particular phase of construction, such as the number of bricks or the number and sizes of windows. Also called a schedule of materials, it can often be prepared by a lumberyard or other supplier.

Test Boring. Sample of the soil upon which a structure is to be built to determine what weight the soil is capable of carrying.

Title Insurance. An insurance policy that protects the owner of real estate from any loss due to defects in his title.

Topographical Plat (topo). A drawing showing the surface features of property.

Transit. A surveying instrument used to measure horizontal angles, levelness, and vertical depth.

Trim out. The stage of construction when final trim items are installed — toilets, moldings, light fixtures, and so on.

Unsecured Loan. A loan in which no material possessions are pledged as security for repayment.

Vapor Barrier. A thin membrane impervious to moisture.

Waterproofing. Making a foundation impervious to water.

English/Spanish Glossary

English	Spanish
angle	ángulo (AHN-goo-loe)
area	área (AH-reh-ah)
basement	sótano (SO-tah-noe)
bathroom	baño (bahn-NYOE)
bathtub	tina (TEE-nah)
bedroom	recámara (reh-CAM-ah-rah)
brick	ladrillo (lah-DREE-yoe)
bring	llevar (yeh-BAR)
cabinet	armario (ahr-mah-REE-yoe)
carpenter	carpintero (car-peen-TEHR-oe)
caulk	masilla de calafatear (mah-see-yah deh cah-lah-fah-teh-AR)
ceiling	cielo raso (see-EH-loe RAH-soe)
cement	hormigón (hor-mee-GONE)
center	centro (CEN-troe)
connect	conectar (coe-NECK-tar)
corner	esquina (es-SKEE-nah)
counter	encimera (en-see-MEHR-ah)
cover (verb)	cubrir (coo-BREER)
cut (verb)	cortar (cor-TAR)
deck	terraza (tehr-RAH-sah)
dig (verb)	cavar (cah-VAR)
dig (noun)	excavación (eks-cah-vah-see-OWN)
dining room	comedor (coe-meh-DOOR)
disconnect	desconectar (des-con-ek-TAR)
door	puerta (PWAIR-tah)
drain	drenaje (drehn-AH-hay)
driveway	camino de entrada (cah-MEE-noe deh en-TRAH-dah)
drywall	tabla-roca (TAH-blah-ROE-cah)
electricity	electricidad (ee-lek-trees-ee-DAHD)

English

Spanish

English	Spanish
exterior	exterior (eks-tehr-EEOR)
fan	ventilador (behn-tee-lah-DOOR)
faucet	llave (YAH-beh)
feet (measurement)	pie (pee-yeh)
finish (verb)	acabar (ah-kah-BAR)
finish (noun)	acabado (ah-kah-BAH-doe)
floor	piso (PEE-soe)
footings	pies (PEE-yehs)
form	formar (for-MAR)
foundation	cimiento (see-mee-EN-toe)
frame	armazón (ar-mah-SONE)
framing	armazonado (ar-mah-sone-AD-oe)
garage	cochera (koe-CHEHR-ah)
glass	vidrio (BEE-dree-oe)
glue (verb)	pegar (peh-GAR)
glue (noun)	pegamento (peg-ah-MEN-toe)
glue (noun)	cola (KOE-lah)
gravel	grava (GRAH-bah)
hallway	pasillo (pah-SEE-yoe)
hang	colocar (koe-loe-CAR)
hardhat	casco (CAHS-coe)
here	aquí (ah-KEY)
house	casa (CAH-sah)
inches	pulgadas (pool-GAH-dahs)
insulation	aislamiento (ayees-lah-mee-EN-toe)
jack (phone)	enchufe (en-CHOO-fay)
jackhammer	martillo perforador (mahr-TEE-yoe per-for-ah-DOOR)
jamb	jamba (JAHM-bah)
joint	junta (WHOON-tah)
kitchen	cocina (koe-SEE-nah)
ladder	escalera (es-kah-LEHR-ah)
laundry room	cuarto de lavar (KWAR-toe deh lah-BAR)
level (verb)	nivelar (nee-beh-LAR)
level (noun)	nivel (nee-BEL)

English	Spanish
lift (verb)	levantar (leh-bahn-TAR)
living room	sala (SAH-lah)
lock (verb)	cerrar con llave (sehr-AHR cone YAH-beh)
lock (noun)	cerradura (sehr-rah-DOO-rah)
long	largo (LAR-goe)
lumber	madera (mah-DEHR-ah)
mark (verb)	marcar (mar-CAR)
mark (noun)	marca (MAR-cah)
measure	medir (meh-DEER)
mortar	mortero (mor-TEHR-oe)
nail (verb)	clavar (klah-BAR)
nail (noun)	clavo (KLAH-boe)
narrow	angosto (ahn-GOES-toe)
need (verb)	necesitar (nes-es-eet-AHR)
need (noun)	necesidad (nes-es-ee-DAHD)
new	nuevo (new-AY-boe)
open	abrir (a-BREER)
paint (verb)	pintar (PEEN-tar)
paint (noun)	pintura (peen-TOR-ah)
paintbrush	brocha (BROE-chah)
painter	pintor (peen-TOR)
parallel	paralelo (pah-rah-LEH-loe)
patio	patio (PAH-tee-oe)
pavement	pavimento (pah-bee-MEN-toe)
pipe	tubo (TOO-boe)
plans	planes (PLAHN-ehs)
plaster	yeso (YEH-soe)
plastic	plástico (PLAH-stee-coe)
please	por favor (poor-fah-VOOR)
plumber	fontanero (fon-tahn-EHR-oe)
plumbing	fontanería (fon-tahn-eh-REE-ah)
plywood	madera contrachapada (mah-DEHR-ah con-trah-chah-PAH-dah)
post	poste (POE-steh)

English

Spanish

English	Spanish
putty	masilla (mah-SEE-yah)
rafter	cabrio (cah-BREE-yoe)
retaining wall	muro de contención (MUR-oe deh con-ten-see-OHN)
roof	techo (TEH-choe)
room	cuarto (KWAR-toe)
sand (verb)	lijar (lee-HAR)
sand (noun)	lija (LEE-hah)
sandpaper	papel de lija (pah-PELL deh LEE-hah)
saw (verb)	aserrar (ah-seer-RAH)
saw (noun)	sierra (see-EHR-rah)
scaffold	andamio (an-dahm-EE-oe)
screen	tela metálica (TEH-lah met-AHL-ee-cah)
screw (verb)	atornillar (a-tor-nee-YAHR)
screw (noun)	tornillo (tor-NEE-oe)
sewer	alcantarilla (ahl-kahn-tar-REE-yah)
shut	cerrar (sehr-RAR)
slab	losa (LOE-sae)
splice (verb)	empalmar (em-pahl-MAR)
splice (noun)	empalme (em-PAHL-meh)
square (verb)	escuadrar (es-kwa-DRAHR)
stake (verb)	estacar (es-tah-CAR)
stairs	escaleras (es-kah-LEHR-ahs)
stake (noun)	estaca (es-TAH-cah)
staple (verb)	grapar (grah-PAR)
steel	acero (ah-SEHR-oe)
straight	recto (REK-toe)
stucco	estuco (es-TOOK-oe)
tape measure	cinta métrica (SEEN-tah MEH-tree-cah)
tar	brea (BREH-ah)
telephone	teléfono (tel-EF-oe-noe)
that's all	eso es todo (ES-oe es TOE-doe)
tile	azulejo (ah-soo-LEH-hoe)
time	tiempo (tee-EM-poe)
tools	herramientas (air-ah-mee-EN-tahs)

English	Spanish
trash	basura (bah-SOOR-ah)
trim (verb)	acabar, terminar (ah-kah-BAR, ter-mee-NAR)
trim (noun)	guarnición, vestidura, chambrana (gahr-nis-ee-OHN), (bes-tee-DOOR-ah), (chahm-BRAHN-ah)
truck	camión (kah-mee-OHN)
under	bajo (BAH-hoe)
underground	subterráneo (soob-tehr-RAHN-neh-oe)
understand	comprender (comb-pren-DARE)
unload	descargar (des-cahr-GAHR)
up, above	arriba (ah-REE-bah)
wall	muro, pared (MUR-oe), (pah-RED)
wash (verb)	lavar (lah-BAR)
watch out!	¡cuidado! (kwee-DAHD-oe)
water (verb)	regar (reh-GAR)
water (noun)	agua (AH-gwah)
wheelbarrow	carretilla (car-reh-TEE-yah)
when	cuando (KWAHN-doe)
where	dónde (DON-day)
who	quién (kee-EN)
why	por qué (poor-KAY)
window	ventana (ben-TAHN-ah)
wire	alambre (a-LAHM-breh)
work	trabajo (trah-BAH-hoe)
worker	trabajador (trah-bah-hah-DOR)
wrench	llave de tuercas (YAH-beh deh TWER-kahs)

Useful Phrases

English: How many?
Spanish: ¿Cuántos? (KWAHN-toes)

English: How much?
Spanish: ¿Cuánto? (KWAHN-toe)

English: How much do you need?
Spanish: ¿Cuánto necesita? (KWAHN-toe nes-es-EET-ah)

English: The job pays___ (an hour, day, per job).
Spanish: Por este trabajo se paga ___ (por hora, por día o por
trabajo). (poor ES-teh trah-BAH-hoe seh paa-GAA) ___
(poor or-ah poor DEE-ah, poor tra-BAH-hoe)

English: I will pay you on___.
Spanish: Le pagaré el _____. (leh pah-gah-RAY el)

English: Start the job_____ (today, tomorrow).
Spanish: Comience el trabajo _____ (hoy, mañana).
(coe-mee-ENS-eh el trah-BAH-hoe oy mahn-YAHN-ah)

English: Monday, Tuesday, Wednesday, Thursday,
Friday, Saturday, Sunday.
Spanish: lunes (LOO-nehs), martes (MAR-tehs),
miércoles (mee-EHR-coe-lehs), jueves (WHEY-vehs),
viernes, (bee-EHR-nehs), sábado (SAH-bah-doe),
domingo (doe-MING-oe)

English: Thank you.
Spanish: Gracias. (GRAHS-ee-ahs)

English: You are welcome.
Spanish: De nada. (DEH NAH-dah)

English: Well done.
Spanish: Buen trabajo. (BWEN trah-BAH-hoe)

English: What is your name?
Spanish: ¿Cómo se llama? (KOE-moe say YAHM-ah)

Index

Page numbers in *italics* indicate illustrations.
Page numbers in **bold** indicate tables.

Build Your Home Improvement Skills
with More Books from Storey

Build a Classic Timber-Framed House by Jack A. Sobon
With these step-by-step instructions, you can build your own enduring, affordable, classic timber-framed house. This classic guide covers finding the right building site, selecting the best timber, using economical hand tools, and much more.

Compact Cabins by Gerald Rowan
Simple living in 1,000 square feet or less! With 62 design interpretations for every taste, this fully illustrated guide will inspire your dream cabin. These innovative floor plans are flexible, with modular elements to mix and match.

Install Your Own Solar Panels by Joseph Burdick & Philip Schmidt
These user-friendly instructions and detailed photos will help you save thousands of dollars in solar panel installation costs. From assembling rooftop racking to setting up the electrical connection, this detailed manual follows the same process professional contractors use.

Learn to Timber Frame by Will Beemer
Master traditional timber-framing skills and tools with this unique beginner's guide. Step-by-step instructions and photos show how timber framing differs from other building methods, and the detailed plans for a 12' x 16' structure will get you building.

Micro Living by Derek "Deek" Diedricksen
Get an insider's perspective on what tiny house living is really like, with profiles of 40 houses equipped for full-time living in 400 square feet or less. Real floor plans highlight inventive space-saving features, and residents offer tips on what to consider before building.

Stone Primer by Charles McRaven
Add the elegance of stone to your home inside and out. These how-to techniques and step-by-step projects, along with stonemason profiles and beautiful photography, will kick-start your DIY goals.

Join the conversation. Share your experience with this book, learn more about Storey Publishing's authors, and read original essays and book excerpts at storey.com. Look for our books wherever quality books are sold or call 800-441-5700.